NANOPHYSICS

Ritika Mahalingam

RAJAT PUBLICATIONS
New Delhi

Nanophysics
by *Ritika Mahalingam*

This edition is published by Rajat Publications
4740/23, Ansari Road Daryaganj,
New Delhi - 110002

ISBN: 978-81-7880-370-8

Edition 2025

Printed and bound in India

RAJAT
PUBLICATIONS

4740/23, Ansari Road, Daryaganj,
New Delhi- 110002 (India)
Mob:- +91-9811484973, +91-9811105389
E-mail: rajat_publication@yahoo.com

Preface

This book has been brought out to acquaint the readers with the hype and the rigmarole behind nanophysics. The emphasis of the book is on presenting forth a thorough introductory manual on the subject, detailing its history, the science and the practice, the techniques, the trends and developments, and the future of this science.

The multifold implications of nanophysics have been discussed most comprehensively, as this science holds the key to understanding the future of humanity in terms of scientific developments. The critical insights presented forth in the book are highly informative, and it serves well for all those interested in understanding nanotechnology and nanophysics in all its machinations, and manifestations.

Ritika Mahalingam

Contents

1

Invisible Engineers

INTRODUCTION

The term "nanotechnology" has evolved over the years via terminology drift to mean "anything smaller than microtechnology," such as nano powders, and other things that are nanoscale in size, but not referring to *mechanisms* that have been purposefully built from nanoscale components.This evolved version of the term is more properly labeled "nanoscale bulk technology," while the original meaning is now more properly labeled "molecular nanotechnology" (MNT), or "nanoscale engineering," or "molecular mechanics," or "molecular machine systems," or "molecular manufacturing." Recently, the Foresight Institute has suggested an alternate term to represent the original meaning of nanotechnology: zettatechnology.

At the most basic *technical* level, MNT is building, with intent and design, and molecule by molecule, these two things: 1) incredibly advanced and extremely capable nano-scale and micro-scale machines and computers, and 2) ordinary size objects, using other incredibly small machines called assemblers or fabricators (found inside nanofactories). In a nutshell, by taking advantage of quantum-level properties, MNT allows for unprecedented control of the material world, at the nanoscale, providing the means by which systems and materials can be built with exacting specifications and characteristics. MNT represents the state of the art in advances in biology, chemistry, physics, engineering, computer science and mathematics. The major research objectives in MNT are the design, modeling, and fabrication of molecular machines and molecular devices. The emergence of MNT - both infant and mature - has numerous social, legal, cultural, ethical, religious, philosophical and political implications.

At the most basic social level, MNT is going to be responsible for massive changes in the way we live, the way we interact with one another and our environment, and the things we are capable of doing.

At the end of the day, it is not the meaning behind the terms that is important, it is the fact that all the many definitions suggest that we have been and are on a rapidly accelerating technological rollercoaster, and rapid change is the track it rides.

Once MNT develops to the stage where we've built the two most essential machines - called the Universal Assembler and the Nanocomputer - *everything has a near-term possibility of significant change*.

There are the so-called inert gases in the air we breathe. They bear curious Greek names of erudite derivation which mean 'the New', 'the Hidden', 'the Inactive', and 'the Alien'. They are indeed so inert, so satisfied with their condition, that they do not interfere in any chemical reaction, do not combine with any other element, and for precisely this reason have gone undetected for centuries. As late as 1962 a diligent chemist after long and ingenious efforts succeeded in forcing the Alien (xenon) to combine fleetingly with extremely avid and lively fluorine, and the feat seemed so extraordinary that he was given a Nobel prize.

Sodium is a degenerated metal. It is indeed a metal only in the chemical significance of the word, certainly not in that of everyday language. It is neither rigid nor elastic; rather it is soft like wax; it is not shiny or, better, it is shiny only if preserved with maniacal care, since otherwise it reacts in a few instants with air, covering itself with an ugly rough rind: with even greater rapidity it reacts with water, in which it floats, dancing frenetically and developing hydrogen.

If 1 gram of sugar in the platinum crucible to incinerate it on the flame, there rose in the lab's polluted air the domestic and childish smell of burnt sugar, but immediately afterward the flame turned livid and there was a much different smell, metallic, garlicky, inorganic, indeed contra-organic: a chemist without a nose is in for trouble. At this point it is hard to make a mistake: filter the solution, acidify it, take the Kipp, let hydrogen sulphide bubble through. And here is the yellow precipitate of sulphide, it is arsenious anhydride – in short, arsenic.

Elements (supplemented by some unstable, artificial ones) arranged in a pattern that helps chemists make sense of them. The most famous book 'about' chemistry is the one that Italian chemist and writer Primo Levi named after this tabulation of matter's building blocks, and it reinforces the impression that chemistry begins with this irregularly shaped grid of symbols. At school students encouraged to learn mnemonics encoding the elements in the first two rows of the table, which are the most important. For undergraduate chemistry it was required that one could recite the whole thing from memory, to know that iridium lies at the foot of cobalt, that europium is sandwiched between samarium and gadolinium.

But chemistry is only incidentally about the properties of the elements, and the science of molecules can afford to ignore many if not most of them. The Periodic Table really belongs to that realm where chemistry becomes physics, where we must wheel out the algebra and the cosines to explain why atoms of the elements form the particular unions called molecules. The table is one of the most beautiful and profound discoveries of the nineteenth century, but, until quantum mechanics was invented by physicists in the twentieth century, one could look upon it only as a mysterious cipher, a kind of crib sheet that served as an empirical reminder that elements come in families whose members show similar proclivities.

A conventional history of chemistry presents it as a quest to understand matter: to ask, what are things made of? This links chemistry with ancient Greek philosophy, with the attempts of Leucippus and his pupil Democritus to formulate an atomic theory of matter in the fifth and fourth centuries BC. It gives us a narrative that progresses from Empedocles' four elements - earth, air, fire, and water - through to Plato's marriage of elemental theory with atomism, skirting cautiously around the medieval alchemists' belief in the transmutation of the elements and alighting gingerly on the phlogiston theory of the eighteenth century. We see antiquity's four-element scheme crumble before the discovery of new 'irreducible substances', and we see Antoine Lavoisier dismantle phlogiston and replace it with oxygen before losing his head under the guillotine's blade in 1794. John Dalton gives us the modern atomic theory in 1800, the list of elements expands enormously throughout that century, and then Dmitri Mendeleev arranges them into the twin-towered edifice of the Periodic Table. The gaps are gradually filled all the way up to uranium, and Wolfgang Pauli and the other quantum physicists explain the table's shape in the 1920s.

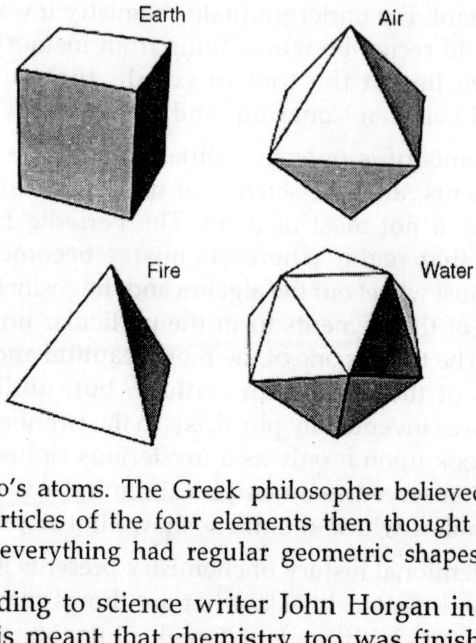

Fig. Plato's atoms. The Greek philosopher believed that the smallest particles of the four elements then thought to comprise everything had regular geometric shapes.

According to science writer John Horgan in *The End of Science,* this meant that chemistry too was finished, once it had the quantum stamp of approval. The implication in several other recent books on the future of science is that the discipline, conspicuous by its absence, has been consumed from both ends. At the most fundamental level, it has become physics (including that immense but overlooked branch called condensed-matter physics, which ponders on how tangible matter behaves). At the most complex level, it is now the domain of biologists, who have expanded their world to embrace the molecular mechanics of the cell.

But these academic turf wars conceal a far more interesting truth. It is a curious fact that many histories of science are written by physicists, who have a tendency to present science as a series of questions posed and then answered. It would be instructive to see the story told instead by an engineer, whose instinct might rather be to ask: what can we make? For, while some of our proto-chemists were wishing to dissect matter, whether physically or metaphysically, others were eagerly rearranging it. This is why the science of molecules is both a creative as well as an analytical pursuit. It has been, at various times in history, concerned with making ceramic pots, dyes and pigments, plastics and other synthetic materials, drugs, protective coatings, electronic components, machines the size of a bacterium. 'What is strange', says chemistry Nobel laureate Roald Hoffmann, 'is that chemists should accept the metaphor of discovery'.

Chemistry is the science of molecules and their transformations. Some of the molecules are indeed *there,* just waiting to be known by us. But so many more molecules of chemistry are made by us, in the laboratory. At the heart of [chemistry] is the molecule that is made, either by a natural process or by a human being. Universities that hide their chemistry departments under the banner of 'molecular sciences' are possibly onto the right idea; for this gently releases the ballast of the Periodic Table and leaves the chemist free to ascend into a world of synthesis, a non-Platonic realm where molecules are designed and made to *do* things, such as cure viral infections or store information or hold bridges together.

As an industrial chemist, Primo Levi moved in this world. He felt a little apologetic about his molecular science: he called it 'a "low" chemistry, almost culinary'. But the power of 'low' chemistry is awesome. It shifts billions of dollars each year, it can make the sick healthy and the healthy sick. Hamburg and Dresden were laid waste by low chemistry, and chemical and biochemical warfare are now more feared in the West than nuclear war. Many people believe that the nuclear bomb was itself the product of physics, but writing $E = mc^2$ does not give you Hiroshima – only separating isotopically distinct molecules of uranium compounds did that.

In *Gravity's Rainbow,* Thomas Pynchon has no doubt where the true power of science lies: the villain of his fantasy from the fag-end of the Second World War is not the Bomb but a new plastic, an 'aromatic heterocyclic polymer' called Imipolex G, developed in a conspiracy between Europe's giant chemicals companies IG Farben, Ciba, Geigy, Shell Oil, and ICI. The message is that 'stuff' speaks louder than theories. After the end of the war, a group from the Allies assembled by Eisenhower claimed that 'Without IG [Farben]'s immense productive facilities, its farreaching research, varied technical experience and overall concentration of economic power, Germany would not have been in the position to start its aggressive war in September 1939.' It was one of IG Farben's subsidiary companies, Degesch, that made the poison gas Zyklon B used in the concentration camps.

Does this mean that molecular science is bad? Of course not – it means that it is a craft full of possibilities. Wonderful, inspiring, inventive possibilities. Terrible, nightmarish possibilities. Mundane but useful things, bizarre things, hard-to-understand things. Molecular science might one day help people to grow a new liver. Raphael, Rubens, and Renoir painted with molecules. Molecules orchestrated the origin of life.

SYNTHESIS: THOMAS PYNCHON'S GRAVITY'S RAINBOW

The origins of Imipolex G are traceable back to early research done at du Pont. Plasticity has its grand traditions and main stream, which happens to flow by way of du Pont and their famous employee Carothers, known as the Great Synthesist. His classic study of large molecules spanned the decade of the twenties and brought us directly to nylon, which is not only a delight to the fetishist and a convenience to the armed insurgent, but was also, at the time and well within the System, an announcement of Plasticity's central canon: that chemists were no longer to be at the mercy of Nature.

They could decide now what properties they wanted a molecule to have, and then go ahead and build it. A desired monomer of high molecular weight could be synthesized to order, bent into its heterocyclic ring, clasped, and strung in a chain along with the more 'natural' benzene or aromatic rings. Such chains would be known as 'aromatic heterocyclic polymers'. One hypothetical chain that Jamf came up with, just before the war, was later modified into Imipolex G.

WHAT ARE MOLECULES?

It is not true that molecule make up every thing. All matter outside of some strange astrophysical environments is made up of atoms; but atoms do not always organize themselves into molecules. Most atoms on their own are highly reactive. They have a predisposition to join up with other atoms. Molecules are collectives of atoms, firmly welded together into assemblies that may contain anything up to many millions of them. But there is a further, subtle distinction to be made. Flann O'Brien's Sergeant Fottrell speaks of 'mollycules' of rock and of iron. Strictly speaking, there are no such things – at least, not in a block of everyday rock or iron. By molecules, we generally mean assemblies of a discrete, countable number of atoms. In the water molecule there are three atoms: two of hydrogen and one of oxygen. A glass of water contains trillions upon trillions of atoms, but a snapshot of the liquid – were it able to reveal such tiny details – would show that at any instant they are nearly all grouped into these three-atom molecules, like a gigantic crowd holding hands in families of three.

These are not woolly, romantic notions calculated to make science appear friendlier although it will do no harm if they have that effect. Rather, they speak of the increasing awareness of the beautiful sophistication of molecular behaviour, which is generally gregarious and rarely linear. It is with these thoughts in mind that is needed to expand on the use of metaphor in molecular science. We cannot do without it, even at the level of one specialist speaking to another. This is true in many areas of science, but in chemistry more than most. Molecules are anthropomorphized mercilessly, and there need be no apology for that. They are unfamiliar things, these molecules, and we need to find ways of making them less so. The publishers of my book about water rightly insisted that ball-and-stick models of H_2O molecules were anathema to the non-chemist reader, guaranteed to ensure that the book stays on the shelf. The strangeness of molecular structure of water is shown into little demons.

This was harmless. But it was reminded of the dangers at a public lecture attended recently on molecular replication. The first question from the floor was 'Are these molecules conscious?' Given that the speaker was talking about a synthetic molecular system that mimics in a very crude way some of the characteristics of living organisms. This was an understandable enquiry. The answer is 'no', if one wants to retain any meaningful working definition of the slippery concept of consciousness. But, once we start to anthropomorphize, we import a baggage of associations, for better or worse. Many people hate the concept of 'selfish genes' because it carries moral connotations. Richard Dawkins calls it 'poetic science', it means – but the poetry of the mechanism gets besmirched by the unpleasantness, as many see it, of the metaphor. The idea that molecules 'cooperate' and 'communicate' is no basis for a philosophy of nature. But it is reason to suspect that, in molecular science at least, a linear, clockwork world view might in the end leave us like the ancient astronomers interpreting planetary motions from a geocentric perspective: trying to shoehorn the observations into a misconceived framework.

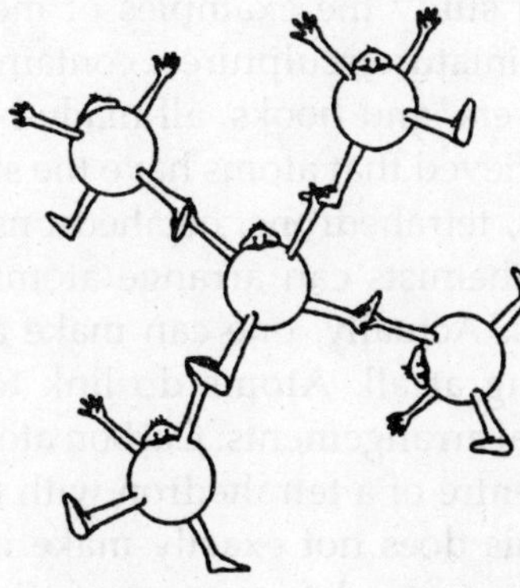

Fig. Making molecules anthropomorphic can help us visualize how they interact.

SHAPE AND SIZE

Primo Levi's *The Monkey's Wrench* is one of the few novels that includes a drawing of a molecule. It is a fearsomely complicated one, and never dream of showing it in a non-technical book about science if my intention was to be instructive. But Levi gets away with it, because he does not want us to understand anything about the molecule, except for one thing: it has a shape and structure. There are some kinds of hexagon in here, and some straight units linking them together. The narrator is talking to a construction worker named Faussone, a man who assembles girders into bridges. He says, the profession studied in school and that has kept alive so far is the profession of a chemist.

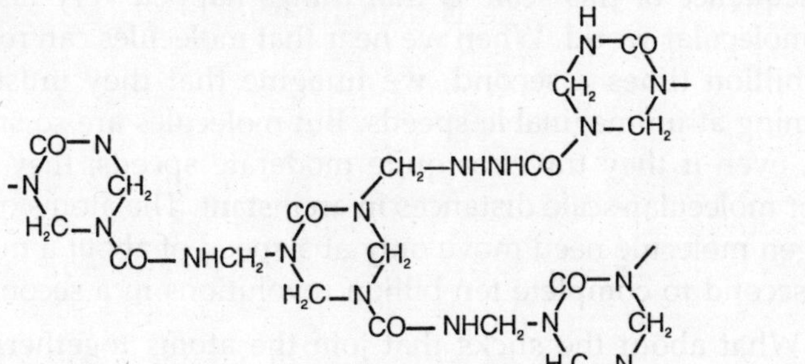

Fig. Primo Levi's molecule.

Now we will study the examples of molecules that can be regarded as miniature sculptures, containers, soccer balls, threads, rings, levers, and hooks, all made by sticking atoms together. Plato believed that atoms have the shapes of 'regular polyhedra': cubes, tetrahedrons, octahedrons, and so on. He was wrong; but chemists can arrange atoms into molecules with these shapes. Actually, one can make a case that Plato was not far wrong at all. Atoms do link together in quite precise geometrical arrangements. Carbon atoms, for example, like to sit at the centre of a tetrahedron with four other atoms at the corners. This does not exactly make it the tetrahedral block that Plato envisaged for atoms of 'fire'; but it shows that Plato's geometric view of the microscopic world held a grain of truth.

How big is this molecule that Levi's narrator draws for Faussone. Each one of those C's, N's, and so forth represents an atom, which is a truly tiny thing. Countless analogies struggle to convey the scale of atoms, but it is not sure that they serve to give an impression any more concrete than that these irreducible particles of the elements are very, very small indeed. For example, if a golf ball were blown up to the size of the Earth, its atoms would be about the size of the original golf gall. Ten million atoms of carbon side by side would make a row about a millimetre long. A small molecule like water is just a few atoms' width in size, about three-tenths of a nanometre. (A nanometre is a millionth of a millimetre.) Primo Levi's molecule is several times bigger. One consequence of this scale is that things happen very fast in the molecular world. When we hear that molecules can rotate ten billion times a second, we imagine that they must be spinning at unimaginable speeds. But molecules are so small that, even if they travel at quite moderate speeds, they can cover molecular-scale distances in an instant. The atoms of an oxygen molecule need move only at a speed of about a metre per second to complete ten billion revolutions in a second.

What about the sticks that join the atoms together? In fact, they take up no space; they are just a convention to help us see what is going on in the diagram. Atoms that are bound together in molecules push right up against one another; in fact, they overlap, rather like two soap bubbles in contact. This is possible because atoms are not like hard billiard balls, but more like rubber balls.

They have a centre that *is* dense and hard, called the nucleus, and this is about ten thousand times smaller than the atom itself - although it is where nearly all of the atom's mass is concentrated. The nucleus has a positive electrical charge. Surrounding it is a cloud of electrons, which are small, light subatomic particles with a negative charge. The electron clouds of two atoms can overlap without danger of electrons colliding, and the two atoms then share some of their electrons: the two clouds merge into one, encompassing both nuclei. When this happens, the two atoms are said to be linked by a *covalent bond.* The sticks in the molecular diagram represent covalent bonds, and they are just a way of helping us to see which atoms are bonded to which.

Here is one of the crucial considerations in talking about molecules, and it is one that complicates the whole attempt: there is no 'best' way of drawing them. One might say: well, never mind the schematic diagrams, why not show what they 'really' look like? But that does not help, because there is no way of taking a photograph of a molecule in the same way as we can photograph a cat or a tree. This is not a matter of technical limitations - it is not that we lack a microscope or a camera capable of resolving such small objects. The fact is the mechanics of seeing make it impossible to 'see' a molecule (or an atom, for that matter) 'as it really is'. The reason is that we see with visible light, which is a wave-like radiation for which the wavelength - the distance between successive crests - varies from about 700 nanometres for red light to 400 nanometres for violet light. In other words, red light fits about 140,000 undulations into a centimetre. This wavelength is hundreds of times larger than a molecule. Roughly speaking, light cannot be focused to a point smaller than its wavelength, which means that objects smaller than that cannot be resolved. No lightbased microscope will ever show us a sharp image of a water molecule.

This is the main reason why people find molecules hard to comprehend, and why diagrams like the one above are a good way to scare readers away from a science book. It seems absurd to be talking in a concrete manner about objects that are not only too tiny to see in practice but too tiny to see in principle. Things that cannot be seen acquire an aura of fantasy, as though they are just a convenient fiction. Molecules are not a fiction, however, and we can prove not only that they are there but that they have definite shapes and sizes.

Some portraits of molecules taken with a special kind of microscope that does not use light to form its images. Beside each snapshot a diagram of the molecule's structure is shown. Well before this type of microscope was invented, the molecules were known to possess these structures; but no one had ever seen them directly. The images are pretty blurry - you would not be able to guess the exact shape of the molecules from these portraits alone. But the shapes seen in the microscope do match up in a convincing way with those expected.

Some of the corroborating evidence is experimental. Even though molecules are too small to be resolved with visible light, they can be 'seen' with radiation of a wavelength comparable to their own size. Radiation with a wavelength of about a tenth of a nanometre corresponds to X-rays, and, by bouncing X-rays off crystals, it is possible to deduce where their constituent atoms are located.

This means that, if a substance can be made in crystalline form, with all its molecules stacked together in an orderly manner, the technique called X-ray crystallography can reveal the structure of the molecules. In principle we should be able to see individual molecules with an X-ray microscope that focuses X-rays just as we focus light in an optical microscope. In practice it is very hard to focus X-rays, although scientists are on the verge of being able to do so. In the meantime we can get by with the electron microscope, in which a beam of electrons is bounced off the sample and focused to make an image. Electrons can act as waves too, and using electron waves we can construct images of very large molecules such as proteins or DNA. These pictures are not detailed enough to show individual atoms, but they do give an impression of the molecules' overall shape.

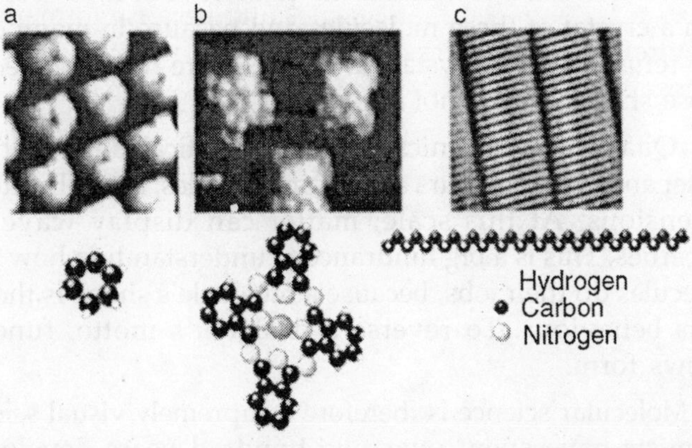

Fig. Molecules 'photographed' with the scanning tunnelling microscope, which is capable of resolving them individually.

Another way to deduce the shapes of molecules is theoretical: it is possible to calculate them. This gets us into the 'algebra' of Flann O'Brien's 'Mollycule Theory', but there is no need to describe it here. It is enough to say that the laws of quantum mechanics enable us to predict how atoms will form bonds and where they will then sit in relation to one another. There is nothing arbitrary about the way that atoms join together.

In particular, atoms of each element have a tendency to form a fixed number of bonds, which is called its valence. Carbon atoms prefer to form four bonds, hydrogen atoms just one. Oxygen atoms form two. The quantum theory of molecular structure is indeed 'a very intricate theorem', and even the best computers can solve the equations only approximately. But it is currently possible to calculate the structures of medium-sized molecules with a fair degree of confidence. Comparison of the predictions with the structures of molecules found by X-ray crystallography typically shows a good match.

Yet there is still no reliable way of predicting the shapes of many of the big molecules found in living cells. In such cases, X-ray crystallography becomes difficult too, both because it is hard to decode the pattern of X-rays scattered from a crystal of these molecules and because in many cases they refuse to form crystals at all. Cells are full of molecules whose shapes we do not know.

Quantum mechanics is a mathematical description of matter and its behaviours at very small scales, typically atomic dimensions. At this scale, matter can display wave-like properties. This is a big hindrance to understanding how life's molecules do their jobs, because a molecule's shape is the key to its behaviour. To reverse a designer's motto, function follows form.

Molecular science is therefore a supremely visual science. Chemists have spent over two hundred years developing pictorial languages to describe their craft, with the result that they now have to be polylingual. There are many different ways to depict molecules, each developed to show the particular aspect that the illustrator wants to emphasize. The English chemist John Dalton began in 1800 by drawing molecules as collections of circular symbols depicting atoms, each embellished with shading or markings to identify the element concerned. It was transparent enough, once you knew the code. Very well – but not easy for the printer, who had to make up these symbols specially. A neat shorthand was to give each element a one or two-letter symbol: C for carbon, O for oxygen, Ca for calcium, Fe for iron. Classically minded even in the nineteenth century, chemists chose to register the metal as *ferrum;* gold and silver, *aurum* and *argentum,* became Au and Ag in the same way. 'Ir' is not iron but the element iridium. But at least the system was *meant* to be self-evident.

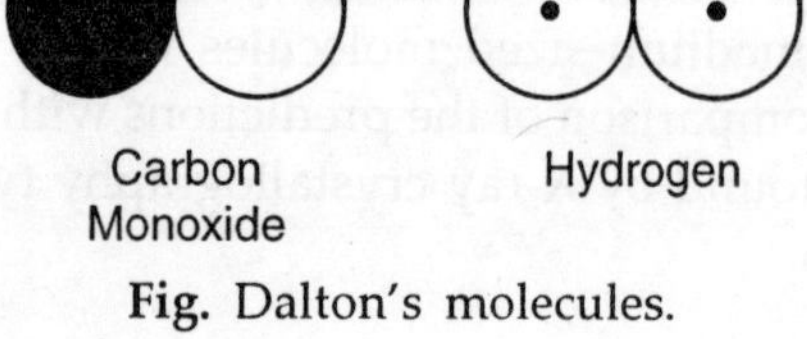

Fig. Dalton's molecules.

Carbon monoxide is simply CO. A multiplicity of atoms of the same element is denoted by a subscript, making the hydrogen molecule H_2. This scheme does not provide a unique inscription for every molecule. Dimethyl ether and ethanol are different substances with different properties, however both can be represented by the *chemical formula* C_2H_6O.

We are back to the lexicological problem: the meaning of a word is determined not only by the letters it contains, but in what order. So there is a need for a formalism that shows how the atoms are linked together. This is where the sticks, representing bonds, come in. The two versions of C_2H_6O, which are called *isomers* (same components, different order), can be represented in this way:

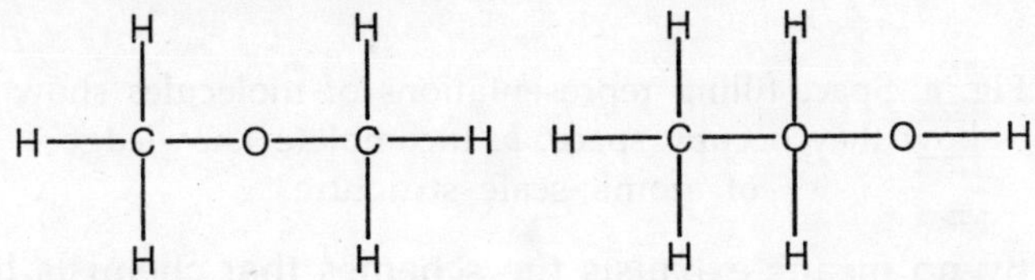

Dimethyl ether Ethanol

A further complication is that the molecule occupies, not the two dimensional plane of a page, but all three dimensions of space.

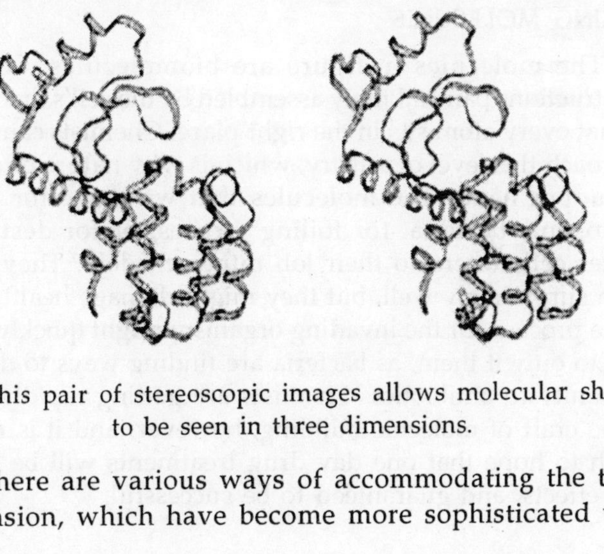

Fig. This pair of stereoscopic images allows molecular shapes to be seen in three dimensions.

There are various ways of accommodating the third dimension, which have become more sophisticated with the advent of computer graphics shows a stereoscopic representation of a medium-sized molecule: with practice we get the two images to overlap and see the shape in 3D.

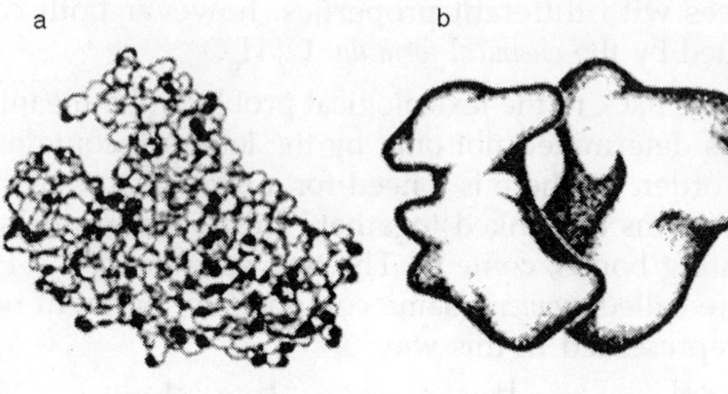

Fig. *a:* Space-filling representations of molecules show how they occupy space. *b:* Incomplete knowledge of atomic-scale structure.

By no means exhausts the schemes that chemists have devised. Sometimes there is a call for 'space-filling models' that reveal how great a volume of space the molecule occupies. Sometimes *ad hoc* schematizations work best, to avoid showing unnecessary detail.

MAKING MOLECULES

The molecules in figure are biomolecules: fantastic constructions painstakingly assembled by the cell's machinery so that every atom sits in the right place. Chemists cannot yet approach this level of artistry, which is why nature often gets the upper hand. The molecules that we make for killing pathogenic bacteria, for foiling viruses, or for destroying cancer cells often do their job rather crudely. They work, often surprisingly well, but they might damage healthy cells in the process. Or the invading organism might quickly find a way to outwit them, as bacteria are finding ways to develop immunity to antibiotics. Chemists are getting rapidly better at the craft of molecule building, however, and it is not too much to hope that one day drug treatments will be free of side effects, and guaranteed to be successful.

Primo Levi put it this way: when you come right down to it, we're bad riggers. We really are like elephants who have been given a closed box containing all the pieces of a watch: we are very strong and patient and we shake the box in every direction and with all our strength. Maybe we even warm it up, because heating is another form of shaking. Well, sometimes, if the watch isn't too complicated, if we keep on shaking, we succeed in getting it together. This is actually something of a worst-case scenario.

Yet, until the last few decades of the twentieth century, the approach that Levi describes, which chemists like to call 'shake and bake', was often the best they could do. Most of the effort in synthetic chemistry is devoted to making so-called organic molecules, which means that they have skeletal frameworks built largely from carbon atoms. Most of the molecules have depicted so far are considered to be organic molecules. In dimethyl ether and ethanol, the carbon framework is rather small. In Primo Levi's molecule, it is a more complicated skeleton. And you can see that nitrogen atoms are also part of his framework, while an oxygen atom serves as a crucial bridge in dimethyl ether. Organic molecules are not necessarily based on skeletons exclusively of carbon – just predominantly so. 'Organic' might seem a strange choice of word, for nearly all of the molecules that organic chemists play with are products not of nature's organisms but of the laboratory.

The term is a historical one, for organic chemistry was once indeed the study of the molecules derived from living organisms. These, it became clear, were largely carbon-based. Why carbon? Atoms of carbon are almost unique amongst the elements in their ability to link together into stable frameworks with complicated shapes: rings, long chains, branching networks. Chemists of the nineteenth century had at best only a dim awareness of how to make new organic molecules. They could modify the molecules that nature provided, chipping fragments off the carbon backbone and replacing them with others; but altering the framework itself was more difficult.

The problem was rendered still more refractory by the fact that they usually had little idea of the true architecture of the molecule they wanted to make. It is a wonder that their shake'n'bake methods got them as far as they did, providing the first synthetic plastics, dyes, and drugs. By starting where some of these chemists started, we can begin to see what molecule building is about, and why it is so difficult - and so desirable. In the mid-1850s, the German chemist August Wilhelm Hofmann, working in London, directed his teenaged student William Perkin to make the compound quinine from the distilled components of coal tar. Quinine is a natural extract of the cinchona tree, and was used to treat malaria. Coal tar was a sticky black residue produced in great quantities by gas works, which sprang up in the early part of the century following the invention of gas lighting. It was unpromising stuff, but Hofmann and others had discovered that by distillation one could separate from it several carbon-rich, odorous ('aromatic') organic compounds such as benzene, toluene, xylene, and phenol. No one knew the structures of any of these compounds - no one could draw stick diagrams like those shown earlier, indicating the connectivity of the atoms. All they knew was how much of each element the compounds contained. Benzene, has the chemical formula C_6H_6, and quinine $C_{20}H_{24}N_2O_2$. The shape of the carbon backbones in these molecules was totally unknown.

Hofmann's procedure was to count atoms. They began with a coal-tar extract that they could convert to a compound called allyltoluidine with most of the right atoms in roughly the right ratios, and hoped that some appropriate treatment of this substance would convert it to quinine. They guessed that two molecules of allyltoluidine (with formula $C_{10}H_{13}N$) might combine with some oxygen and hydrogen to make the drug. But it was a long shot, for there are many ways of linking ten carbon atoms together; and, as it happened, the carbon framework of allyltoluidine is not the same as that of half a quinine molecule. So the experiment, which Perkin conducted in a laboratory rigged up at his parents' house in east London, did not work – it just gave a rust-coloured sludge, something unhappily familiar to organic chemists. Yet the young Perkin did not give up – he tried instead starting with an organic compound called aniline in place of allyltoluidine. This time the sludge was black. But when it was dissolved in methylated spirits, it produced a glorious purple colour, which, to Perkin's delight, would dye silk. He had discovered the first aniline dye. He set up a factory with his brother and father to make the stuff, and it was soon being manufactured in large quantities in Britain and France.

This marked the beginning not only of the synthetic dye industry but of the entire modern chemicals industry – for many of today's chemicals companies, such as BASF, Ciba-Geigy, and Hoescht, began as manufacturers of aniline dyes. By the last quarter of the century, the synthesis of organic molecules had become a less haphazard business. August Friedrich von Kekulé deduced in 1857 that carbon atoms are four-valent – they like to form four bonds. And in 1865 he proposed that benzene, to which all of the aromatic coal-tar molecules were related, contains a *ring* of six carbon atoms – an ubiquitous leitmotif of organic chemistry.

In 1868 the German chemists Carl Graebe and Carl Liebermann synthesized the alizarin molecule, which is responsible for the red colour of the dye extracted from the root of the madder plant. This was one of the most commercially important of all natural dyes, and Graebe and Liebermann's synthesis eventually made it available much more cheaply as an artificial product. The synthesis of alizarin stands as a landmark in molecule making for two reasons. First, it was achieved by planned modification of the starting material (another coal-tar aromatic compound called anthracene), rather than by cooking up the ingredients and hoping for the best. The chemists had some knowledge not only of the formula but also of the chemical structure of anthracene, which they knew to be related to that of alizarin. (In fact they guessed the wrong structure, but fortunately this turned out not to matter.)

Organic chemists call this kind of procedure – in which a starting molecule is converted systematically, bit by bit, to the desired product – a *rational synthesis.* Secondly, by making alizarin in the laboratory Graebe and Liebermann had shown that organic chemistry now had a prowess to rival nature's. It was possible to make the complicated molecules found in living organisms, which chemists today call *natural products.* So was the synthetic red dye made by Graebe and Liebermann, and later by the ton in chemicals factories, identical to the natural red madder dye? Yes and no. The extract traditionally obtained from madder root is a mixture of several different compounds. Alizarin is the main colourant molecule, but the extract also contains a closely related compound called purpurin, which (despite the name) imparts an orange colour. The process that converted anthracene to synthetic alizarin also generated several side products, mostly molecules with structures very similar to alizarin. Perkin was one of the chemists who, in the early 1870s, identified at least four side products in the alizarin manufactured industrially. There were doubtless many others present in even smaller quantities.

So, whereas the alizarin molecules made synthetically were identical to those extracted from madder root, the actual synthetic dyestuff was different from the natural dye – and both were impure. This is true of just about any 'natural-product' compound manufactured industrially, because all the synthetic procedures used by organic chemists generate side products. It does not necessarily mean that synthetic chemicals are better or worse than the equivalents extracted from natural sources: both are likely to be impure to some degree. But chemists place great value on purity, and spend much time ridding their products of impurities. Natural extracts, in contrast, are complex mixtures unless processed to separate the components. The commercial value of substances derived from coal-tar compounds was not restricted to dyes. Paul Ehrlich, a German medical scientist, used the new synthetic dyes in the 1870s for staining cells, making them easier to study under the microscope. He noticed that some dyes would kill the bacterial cells that absorbed them, which suggested therapeutic possibilities. He began to synthesize dye compounds to test as drugs, and in this way he found in 1909 an arsenic-containing dye that would kill the parasite that causes syphilis. As Salvarsan, this drug offered the first relief from the deadly disease since the medieval use of mercury.

It was the beginning of modern chemotherapy. Nineteen years later Alexander Fleming discovered penicillin, a compound produced by a mould, which killed bacteria. This was the first antibiotic, and it revolutionized surgical medicine by greatly reducing the risk of wound infection. Many other natural products have beneficial physiological effects: salicylic acid, an extract of willow bark, has both antiseptic and analgesic (pain-killing) properties, and a closely related molecule provides the drug aspirin, manufactured by Bayer since 1899. Chemists and medical scientists continue to comb through nature's arsenal of molecules for potential drugs, and then to find ways of synthesizing those that work. One of these that has gained fame in recent years is the compound called paclitaxel, better known by the trade name of taxol. It is a natural product of the Pacific yew tree, and was found in the 1980s to be highly effective at preventing cells from dividing.

This makes it a potential anti-cancer agent, since cancer is the result of uncontrolled proliferation of cells. It has been approved by the US Food and Drug Administration for the treatment of breast, lung, ovarian, and prostate cancer. But the Pacific yew does not provide a reliable source, since each tree yields up only a few milligrams of the compound, and it has to be extracted from the bark, so that the tree dies. Already an endangered species, it would be wiped out before meeting the global demand for taxol. There is clearly a call for synthetic taxol. The molecular structure of taxol is fiendishly complicated. Its backbone consists of four rings of carbon: one with four atoms, two with six, and one with eight. Various subsidiary groups of atoms dangle from this framework.

There is no standard chemical reagent available that has this skeletal form – it must be constructed from scratch. This is a profound challenge for the chemical rigger. Primo Levi's character gives an indication of the way that a synthetic organic chemist today would go about it: as you can imagine, it's more reasonable to proceed a bit at a time, first attaching two pieces, then adding a third, and so on.

It takes more patience [than shake'n'bake], but actually you do get there first. And most of the time that's the way we do it. A synthesis like this has to be carefully planned. The most common method of planning is that devised by the American Nobel laureate Elias J. Corey, who called it 'retrosynthetic analysis'.

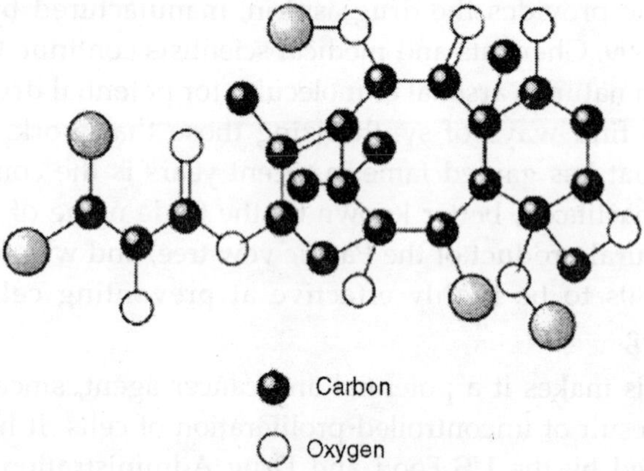

Fig. The taxol molecule. the black spheres denote carbon atoms, and the white spheres oxygen. The larger grey spheres are 'substituent' groups containing carbon, oxygen and hydrogen

As the name implies, you work conceptually backwards from the finished product, as if disassembling a model of the molecule. At each step you break bonds that you can see how to forge, so that, when it comes to conducting the forward process, you have already seen how each link is to be made. The trick is to work back to starting materials – fragments of the carbon framework – that are readily available, or that can easily be synthesized from off-the-shelf compounds. In the case of taxol, two groups 'got there first'. A team at the Scripps Research Institute in California, led by K. C. Nicolaou, and a group working under Robert Holton at Florida State University, described multi-step synthetic procedures within a week of one another in 1994.

There is no unique way to make a molecule this complex, nor indeed a 'best' way – several alternative pathways have since been reported. But they all remain too complex to be viable for large-scale production, and taxol is currently manufactured 'semi-synthetically': from an intermediate compound found in the yew needles, which is like half-built taxol. The synthesis can be completed relatively efficiently in the laboratory, and the needles can be removed without killing the trees.

Fig. A molecular 'big wheel' Each pyramid is a cluster of molybdenum and oxygen atoms.

As described here about making organic molecules, but we should emphasize that many chemists build molecules based on elements other than carbon. These are often rather small molecules, since other elements do not so readily form the large, intricate frameworks that carbon will adopt. It was made by Achim Müller's group at the University of Bielefeld in Germany, and measures fully four nanometres across (that is, about fifteen times the width of a water molecule, and tens of thousand times smaller than the width of a human hair). When metals and oxygen combine, they do not usually stop in the middle ground of large molecules: either they will form molecules of a few atoms each or they will crystallize into mineral-like solids ('infinite molecules', if you like). Chemists have recently become very interested in large *inorganic* molecules like the ones shown here, because they can show unusual and possibly useful behaviour such as magnetism or electrical conduction. Components such as transistors on microchips are made of inorganic materials, primarily silicon and silicon dioxide. Tailor-made components for a molecular-scale electronics technology are just one of the items on the menu of today's sophisticated molecular cookery.

2

Molecules of Life

When scientists and poets reach the same conclusions. Pondering the question 'What is life?

Life is a pattern of chemical processes. This pattern has special properties. It begets a similar pattern, as a flame does, but it regulates itself as a flame does not. So when we have said that life is a pattern of chemical processes, we have said something true and important. But to suppose that one can describe life fully on these lines is to attempt to reduce it to mechanism, which is believed to be impossible.

Is life mere molecules, acting together with awesome but in principle explicable complexity? Or is something more involved? We simply do not know. Scientists take a bottom-up approach, assuming the minimum and invoking only testable hypotheses. Whether this will eventually lead to a point beyond which science is powerless to proceed, we cannot yet say. But no such point has so far become apparent. It seems possible that life – which we might loosely define as an organism that can reproduce, and respond to and extract sustenance from its environment – may be nothing but molecules and their relationships. Indeed, this seems extremely likely. It need not be disappointing; quite the contrary, it would be remarkable. That a conspiracy of molecules might have created *King Lear* is a possibility that makes the world seem an enchanted place. The human mind will ever be *explained* in molecular terms, any more than *Lear* is explained by the alphabet. Most scientists do not believe so either. Phenomena are hierarchical: all things cannot be understood by considering only what transpires on a single rung. No matter how well this can be understood the way a transistor works. If sown seeds are fail to grow, better to begin by thinking about the nutrient content, humidity, and temperature of my soil than by performing a genetic analysis of the seeds. Much of the skill in doing science resides in knowing where in the hierarchy you are looking - and, as a consequence, what is relevant and what is not. It is worth spelling this out before we explore the molecules of the living world, because a molecular view of biology is often branded as reductionistic - as an attempt to explain every aspect of life at the molecular level of genes.

This is indeed sometimes the best way to proceed, for molecules are after all the smallest functional units on which life is founded. But if we accept, as most scientists do, that by descending the ladder to the microworld we must forgo a whole range of questions and answers about life, then there seems to be nothing obviously objectionable about the descent. Indeed, this path has led us to a clearer understanding of our fundamental nature. Molecular biology has helped to fill the major gap in Charles Darwin's evolutionary theory: the issue of the *mechanism* of natural selection. It has given us at least some inkling of how life came into being on a planet of gas, rock, and water. It has saved lives and relieved much pain and suffering. It has helped us to understand why medicines do not always work as we might hope, why irresponsible use of antibiotics has bred superbugs, how the AIDS virus does its terrible work. The study of life's molecules became the major science of the twentieth century's second half, and looks set to have an ever greater impact on our lives in the future. It is perhaps the one area of science in which some degree of knowledge is no longer a luxury.

THE VITAL FORCE

Organic chemistry was once considered different from the rest of chemistry. Many scientists believed in the early nineteenth century that organic matter was the product of a vital force operating in living organisms, which the chemist could never mimic in the laboratory.

But by 1818 the influential Swedish chemist Jons Jacob Berzelius saw tautology looming in the idea of vitalism, while despairing of ever getting beyond it: the cause of most phenomena within the Animal Body lies so deeply hidden from our view, that it certainly will never be found. We call this hidden cause *vital power;* and like many others, who before us have in vain directed their deluded attention to this point, we make of us a *word* to which we can affix no idea.

Yet in the same breath, as it were, Berzelius hinted at how to go further: This *power to live* belongs not to the constituent parts of our bodies, nor does it belong in them as an instrument, neither is it a simple power; but the result of the mutual operation of the instruments and rudiments on one another.

Understanding the molecular basis of life is not so much about appreciating what the molecules are, as what they do to one another. The molecular nature of life is not a gallery but a dance. In Haldane's time it was not unusual to regard life as a series of chemical transformations being conducted as if in some vast network of laboratory glassware. The key to it all, scientists believed, was metabolism: how we obtain energy from food. But you will not make an organism by throwing into a pot all of the purified molecular components of the cell.

A modern view of molecular biology is concerned with *organization* in time and space. How do the molecules of life arrange themselves amongst the cell's compartments, how are they shifted around, how do they communicate so as to synchronize their action? We can ask these questions only because we can now inspect the working cell at the molecular level, taking measurements and snapshots of molecules going about their business. And so the cell becomes a community. Yet it is a community of Byzantine complexity. Molecular biology is not difficult in the way that theoretical physics is difficult – the concepts are not unfamiliar, abstract, or mathematically abstruse. The difficulty arises because there is so much going on all at once. We react with surprise and shock when things go wrong with our own molecular machinery, but it is far more astonishing that the machinery works at all. Frequently it does so because it is designed to be robust in the face of the world's vicissitudes.

There are checkpoints, safety mechanisms, back-up plans, careful recordkeeping. No human-made mechanism is anything like as sophisticated or as well organized as a cell. It is important to bear in mind that the cell is a community of automata. Its members have no volition, no foresight, no memory, no altruism (nor selfishness, in the strict sense). They often collaborate so beautifully that it is easy to forget this.

On the other hand, cells can be unpredictable, because we know so little about how they work. They might survive when we expect them to die, or they might react to a potential drug in totally unforeseen ways. Molecular biology works at the level of the cell, and seldom talks about the whole organism. The cell is the 'atom of life' – you cannot get any smaller and still be alive. (Viruses are a debatable exception – they are little more than genes wearing a coat, but they cannot reproduce without hijacking the machinery of the cells they infect.) This need not be as restrictive a viewpoint as it might seem, since we can understand an awful lot of our requirements as humans according to what goes on in a single cell. A human cell needs oxygen and sugar to make new molecules and to replicate itself – so we breathe and eat. Nerve impulses start at the level of the cell. Our tissues – skin, hair, bone, muscle – are put together molecule by molecule in the cell. We excrete to remove the cell's waste. We shiver and sweat to stabilize our cells' temperature. Very many questions about the way we function can, in other words, be addressed on the rung of molecular biology. Of course, amongst those that cannot are many of the most interesting.

THE PLAYERS

A socialist Haldane attributes to Engels the proposition that life is 'the mode of existence of proteins'. This vitalistic statement implies that proteins are inherently alive, an idea that Haldane squashes. But he has no objection to the idea that *proteins* are the stuff of life.

Proteins are substances found ubiquitously in living cells. Many of them are enzymes, molecules that catalyse processes of chemical change. Enzymes speed up chemical reactions by factors of perhaps several millionfold and thereby ensure that the body's chemistry is not impossibly slow. They were discovered from studies of fermentation: *enzyme* is Greek for 'in yeast'. In the late nineteenth century it was found that enzymes could be extracted from yeast cells and purified, yet would remain capable of bringing about fermentation despite the fact that they were no longer part of a living system.

This discovery helped to establish that the chemistry of life works according to the same principles as the chemistry of non-living matter. If the cell is a city, enzymes are the workers. To keep the city running, raw materials are imported and converted into useful items. Enzymes populate the cellular factories in which this is done. One curious aspect of this manufacturing industry is that it includes factories for making the workers themselves: enzymes too are put together on a production line. Not all proteins are enzymes. Some serve a structural role, providing the tissues of the body. Some act as the cell's police force, others carry packages to and fro in a protein shuttle service running on protein tracks. Some operate the portals to the cell, sitting in the outer membrane and opening or closing in obedient response to the instructions they receive.

There are something like 60,000 different varieties of protein molecule in human cells, each conducting a highly specialized task. It would generally be impossible to guess what this task is merely by looking at a protein. They are undistinguished in appearance, mostly globular in shape and composed primarily of carbon, hydrogen, nitrogen, oxygen, and a little sulphur. All the proteins that fulfil a particular task have the same shape and structure - the seemingly amorphous blob is in fact exquisitely designed and assembled.

Many enzymes are shaped a little like a knobbly kidney bean, with a cleft in the inner curve. This cleft is where the action is - where the molecule performs its catalysis. Some proteins do their jobs in groups: they become 'subunits' of a many-protein assembly. The enzyme tryptophan synthase, which bacteria possess, is one of these, built from four detachable subunits.

This enzyme synthesizes the small molecule tryptophan, which is essential to all organisms. Humans do not possess the enzyme, and so we have to get our tryptophan ready-made by eating organisms that have constructed it. As this example implies, enzymes and other proteins are commonly given names that reveal their function. Alcohol dehydrogenase is an enzyme that takes a hydrogen atom from ('dehydrogenates') an alcohol molecule. ATP synthase synthesizes the molecule ATP.

But not all protein names are so transparent. Haemoglobin, which carries oxygen in the bloodstream, gets its name from the Greek for blood (*haeme*) along with the fact that it is globular. Myoglobin, to which haemoglobin donates its oxygen cargo in muscle tissue, derives the first part of its name from the Greek word for muscle. Other names are more whimsical. Elastin is an elastic protein found in many flexible body tissues, such as blood vessels and vocal cords. Ubiquitin is a protein found just about everywhere in the body, because it plays a central role in the universally necessary process of destroying obsolete proteins.

A protein molecule is in fact a single chain of small molecules linked together. The chain is folded and coiled on itself so densely that it looks like just a mass of atoms. But close inspection of the structure obtained from X-ray crystallography allows us to follow the strand as it twists and turns through the compact globule. Protein chemists sometimes show this strand explicitly in a different kind of representation of a protein. Here you can see that the structure is built up from certain repeating features or 'motifs', such as coils, which are called alpha helices, and so-called beta sheets, where several parts of the strand lie parallel to one another.

Fig. In this representation of a protein molecule, its folded-chain structure is made explicit. The flat, roughly parallel arrows denote beta-sheet motifs, while the coiled segments bearing chevrons are alpha helices. The molecule is the enzyme phosphotyrosine protein phosphatase from a cow.

A protein's structure can be conceptually decomposed even further. The chain is made up of small characteristic clusters of atoms, joined in a sequence like beads on a string. These clusters were once separate molecules, called *amino acids*. There are twenty varieties of amino acids in natural proteins. In the chain, one amino acid is linked to the next via a covalent bond called a *peptide bond*. Both molecules shed a few extraneous atoms to make this linkage, and the remainder – another link in the chain – is called a *residue*. The chain itself is termed a *polypeptide*.Any string of amino acid residues is a polypeptide.

We can make them ourselves simply by heating up a mixture of amino acids. But we will not make a protein this way. In a protein the order of amino acids along the chain – the *sequence* – is not arbitrary. It is selected (that is, naturally selected, in Darwin's sense) to ensure that the chain will collapse and curl up in water into the precisely determined globular form of the protein, with all parts of the chain in the right place. This shape can be destroyed by warming the protein, a process called denaturation. But many proteins will fold up again spontaneously into the same globular structure when cooled.

In other words, the chain has a kind of memory of its folded shape. The details of this *folding process* are still not fully understood - it is, in fact, one of the central unsolved puzzles of molecular biology. We do know, however, what holds the polypeptide chain in its compact form in a protein molecule. Many parts of the chain are able to form weak bonds with each other, called *hydrogen bonds.* These glue the chain into alpha helices and beta sheets, for example. And some parts of the chain are held together by stronger bonds forged between sulphur atoms, which dangle from the residues of the amino acid cysteine. Some residues are relatively insoluble in water, and these will tend to gather together in the core of the protein globule, surrounded by more water-soluble parts of the chain. So the folded structure that results depends on the character of the various residues, and where they are located along the chain - in other words, on the sequence. You could say that protein molecules are manufactured with their own folding instructions.

How does the cell's machinery 'know', when making a protein, in which order the amino acids should be strung together? This is where DNA (deoxyribonucleic acid) enters the picture. The sequence of every protein in the body is encoded in the DNA molecules that reside in every cell. Proteins do all the work (or most of it); DNA sits passively waiting to be read, when the need for a protein arises. DNA information is written in a different language from protein information, but the cell can translate between the two. DNA is another string of small molecular units - another *polymer.* But its building blocks are different. Rather than amino acids, they are molecules called *nucleotides.* DNA is a library of protein structures, written in the characters of the *nucleotide sequence.* The information for making each protein is, roughly speaking, encoded in a sketch of DNA called a *gene.* There is clearly some impressive coordination involved here.

Whenever a job needs doing by enzymes, the message must get through to the region where the DNA resides. In human cells and those of all other organisms other than the most 'primitive' singlecelled bacteria, this is a central compartment called the *nucleus,* which is fenced off within its own membrane. Organisms with cell nuclei are called *eukaryotes.* In human cells DNA is packaged up in bundles called *chromosomes.* To make a protein, the stretch of DNA containing the corresponding gene is unravelled and read. In fact, proteins are made not in the nucleus but in a different compartment called the endoplasmic reticulum, a labyrinthine network of membrane channels. The gene is *transcribed* first into a molecule related to DNA, called RNA (ribonucleic acid). The RNA molecules travel from the nucleus to the endoplasmic reticulum, where they are *translated* to proteins.

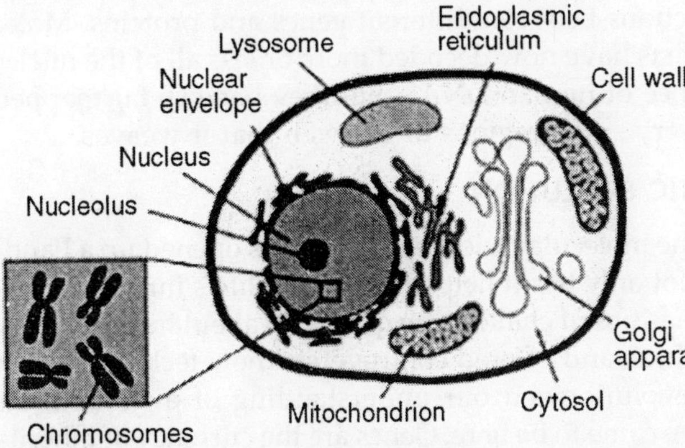

Fig. Cells of humans and other eukaryotic organisms sequester their genetic material (DNA) in a central nucleus. Various other compartments (organelles) perform a variety of other functions, such as protein synthesis and energy production. Note that the DNA is bundled into chromosomes (as shown here) only when the cell is about to divide; at other times it is unravelled into thin strands.

The proteins are then shipped off to where they are needed. So the cell's molecules must be capable of communication and transport. The whole affair is regulated so that proteins are not being made willy-nilly, but only on demand. If the cell was constantly making all the proteins at its disposal, it would rapidly seize up. A key insight into how the cell maintains order amongst its component parts emerged from the work of French biochemists François Jacob and Jacques Monod in the 1960s. They showed that genes regulate one another, switching each other on and off via the agency of the proteins they encode. For example, some genes that encode proteins used in the cell called *structural genes* are partnered by *regulatory genes* that encode repressor proteins. When the regulatory gene is switched on, the repressor protein is synthesized and binds to the structural gene, preventing it from being 'expressed' (transcribed and translated into its protein). Jacob and Monod called these regulated stretches of DNA *operons*. They provide just one illustration that the cell employs a weblike network of interactions between different genes and proteins. Molecular biologists have now decoded more or less all of the nucleotide sequence of human DNA – but they have so far mapped out only very small regions of the web that it weaves.

GENETIC EVOLUTION

The molecular science of genes has opened up a Pandora's box. Not only has it helped to explain life's innermost riddles, but it has posed challenging questions about human behaviour and ethics, and offered controversial new technologies. It has also revolutionized our understanding of evolution, and of how we came to be here. Genes are the currency of inheritance: they are an inevitable bequest from our parents. The idea that characteristics are passed on from parent to offspring is a very old and obvious one, but it was made more concrete by the work of the Austrian priest and biologist Gregor Johann Mendel in the nineteenth century. His experiments on the propagation of peas led him to suppose that there are 'particulate factors' that mediate heredity, passing from the cells of the progenitors to their progeny.

It soon became clear that these 'factors', later called genes, were molecular in nature, but throughout the first half of the twentieth century many scientists thought they were protein molecules. Haldane, as we have seen, shared this belief. Not until Francis Crick and James Watson deduced the structure of DNA in 1953 was there a compelling case for regarding DNA, rather than proteins, as the molecular stuff of heredity, the fabric of genes. This placed evolution on a concrete molecular basis - for what a fertilized egg gets from its parents is not a preformed body but a set of genetic instructions for a body plan. The evolutionary changes that happen slowly from generation to generation are due to changes in the molecular make-up of the genes. DNA is copied when a cell divides - but not always perfectly. So the DNA that a child gets from mother and father may be a slightly flawed amalgam of both their genes. Generally these flaws will not matter. Sometimes they will be harmful (but note that most genetically based diseases are the result of a child *inheriting* a faulty gene, not acquiring one from random copying errors).

Very rarely, a genetic mutation will have a beneficial effect, making the organism better equipped for survival. The advantage might be extremely slight - but evolution advances through such infinitesimal steps, as tiny advantages lead to fractionally higher reproductive success and thus to a slow increase in the incidence of the mutated gene in the population. What this all means is that genes are a molecular record of evolution. The common ancestor of humans and rabbits shared the same set of genes. The differences that now exist between the respective total complement of genes (the *genomes*) of humans and rabbits reflect the divergence attributable to an accumulation of genetic mutations. This enables scientists to reconstruct evolutionary histories - to deduce the order in which species diverged - from the molecular structure of genes.

Previously, palaeontologists had to make such deductions on the basis of the shapes of bodies or bones; now they have a more readily quantifiable molecular measure of evolutionary change. In particular, evolutionary trees or *phylogenies* can be reconstructed by comparing the special DNA that is housed in a compartment of cells called the mitochondrion. This is the cell's furnace, where the energy is produced. Unlike the DNA in the nucleus, mitochondrial DNA comes directly from the mother. Whereas nuclear DNA is changed by reshuffling the genes of both parents, mitochondrial DNA is changed only by the gradual accumulation of mutations from generation to generation - so it is a better record of evolutionary change. In 1987 Allan Wilson of the University of California at Berkeley and coworkers compared the mitochondrial DNA of people from many racial groups. Knowing the average mutation rate, they figured out that all of the samples, and by extension the mitochondrial DNA in all living humans, derived from a single version that existed 200,000 years ago in the cells of an African woman - the common ancestor of all of humanity. Molecules contain a record of history richer than anything to be found in fragments of clay pots or ancient burial mounds.

THE RNA WORLD

All living organisms, from the humblest of bacteria to the most regal of kings and queens, have their genetic material packaged in DNA, and put into effect by proteins. This implies that all life has a common origin. The most primitive single-celled organisms must have contained proteins and DNA very similar to those in 'simple' bacteria today. But what came before? The molecular symbiosis whereby DNA encodes proteins and proteins help DNA to function and to replicate is fantastically sophisticated even in a bacterium. Neither proteins nor DNA could have come spontaneously into existence from fragments of organic molecules scattered throughout the seas and lagoons of the early Earth: their structures are just too complex to have assembled at random.

It is easier to understand (at least in principle) the 3.8 billion years of evolution from the earliest bacteria or algae to the present day than to understand how, over maybe just a few hundred thousand years, Earth changed from a barren planet to one that cradled life. Chemists have devised many inventive schemes according to which the inorganic constituents of the young Earth, such as methane, carbon dioxide, ammonia, water, and nitrogen, might have become transformed into the amino acids and sugars needed to make the molecules of life.

They are all tentative; no theory yet prevails for the chemical origin of life. But the conceptual difficulties in progressing from rock, gas, and water to prototypes of biomolecules are still smaller than those of turning these building blocks into functioning cells full of proteins and DNA. It is a chicken-and-egg problem: on their own, both proteins and DNA are useless. The favourite way out of the conundrum is to switch attention to the humble go-between: RNA, which carries the genetic information to the machinery of protein synthesis. RNA is much more versatile than DNA. In the 1980s, RNA molecules in the cell were discovered that could act as catalysts for their own rearrangement. Human genes are corrupted with a lot of 'junk' that needs to be excised before the message can be clearly read.

This junk gets copied into RNA, but is then snipped out before the RNA is translated into a protein. This editing is largely conducted by enzymes; but some RNA molecules can do it unassisted. These are called *ribozymes,* reflecting the fact that they show enzyme-like tendencies. During the 1990s, biochemists greatly expanded their appreciation of RNA's abilities. Using the techniques of biotechnology developed for manipulating and rewriting DNA, they have made synthetic RNA molecules that can conduct all manner of chemical processes, such as linking together nucleotides or forming bonds between carbon atoms.

These studies show that in principle RNA is versatile enough to bring about many of the chemical transformations that would have been necessary for life to begin. In short, RNA can act both as a gene-carrier and as a worker. Many scientists researching into the origin of life therefore postulate an era that they call the *RNA world,* which existed before the double act of proteins and DNA appeared. RNA is still a fiendishly difficult molecule to make under conditions comparable to those on the early Earth. But the RNA world breaks the impasse posed by the mutual dependence of proteins and DNA, and so provides a conceptual link between the formation of small organic molecules and the appearance of the first primitive cells.

We simply do not know whether the DNA-protein partnership is the *sine qua non* of life, and it would be rash to suggest that life can have no other molecular basis. But none has been found. It is not so difficult to imagine that slight, systematic modifications to DNA could give rise to an alternative genetic system – but no organisms show such a thing.

SYNTHETIC LIFE

If we do eventually come to understand the origin of life – not necessarily how it really began, but at least how it *might* have begun – could we then rerun it in the laboratory? Can we create life from scratch?

Enough is known already about the molecular basis of life for researchers to be able to speculate about building an artificial cell. It sounds perhaps like a frightening prospect. What if we happened to make a cell that was far better at replicating than 'natural' cells? Would they colonize the planet like an alien invasion?

This is not science fiction. In fact, a synthetic cell will be made within the twenty-first century, for better or worse. It is already routine to make synthetic DNA and to rewrite genes, and many chemists are working on building 'designer proteins' from scratch. The biochemists Jack Szostak, David Bartel, and Pier Luigi Luisi have proposed that:

> Advances in directed evolution and membrane biophysics make the synthesis of simple living cells, if not yet foreseeable reality, an imaginable goal.

They suggest that a 'minimal cell' could be constructed from tailormade ribozymes. A primitive version of a ribozyme that can assemble RNA (and so potentially replicate itself) has already been reported. These could be encapsulated within artificial membranes like those of cells, but with an ability to grow and divide: Luisi has made such 'replicating membranes'. The replicating 'protocells' might evolve RNA molecules capable of assembling amino acids into proteins. This would then allow us, say the researchers, to 'replay the tape of early evolution'. Those who foresee terrible purposes in such experiments might bear in mind that this is an absurdly difficult way to try to develop some deadly weapon, when lethal chemical and biological weapons can be made already with relative ease. All the same, it is impossible to know where such research might lead. That is the reality of molecular science: it is a creative discipline, which gives us something to show for our efforts at the end. Therein lies all the artistry, all the wonder, and all the peril. In the end, we will only get the molecules we deserve.

MOLECULES MATERIALS

The hardest part of space travel apart from the boredom and the danger is the leaving. In the vacuum of space, free from strong gravitational influences, a small burst of propulsion will keep a rocket moving almost indefinitely. So most of the fuel that a rocket takes on board is needed simply to escape the Earth's gravity. This fuel and the engines that burn it account for the greater part of a rocket's mass. Recall vividly the Apollo missions that left the Earth as a gleaming tower and returned as a tiny nub from the nose. If we could launch spacecraft from outside Earth's atmosphere, the payload would therefore be much diminished. In his novel *The Fountains of Paradise,* Arthur C. Clarke suggested how this might be done. He posited the Space Elevator: a platform positioned in geostationary orbit around the Earth, tethered to the ground by a long, superstrong cable. Space hardware and any passengers are shuttled up by elevator to the platform, from where they can be launched into space with a fraction of the fuel requirements needed for a ground-based takeoff. To tether an orbiting platform to the Earth's surface, we would need cables far stronger than anything currently available.

They would need to be lightweight too – the weight of that much steel cable would be immense. It does not take the speculative concept of a Space Elevator to explain why we need materials that are strong, tough, corrosionresistant, lightweight, and so forth. But scenarios like this serve to motivate the question of just how far one can go in improving materials' properties. Superstrong 'space tethers' are also being considered for launching objects into space by a kind of slingshot process in which the payload is temporarily attached by a thread to an orbiting satellite. Such tethers have to be both lightweight and strong. And there is no lack of more mundane demands for strong cables: for example, to hang bridges and tether drilling rigs to the seabed. Tough fibres have always been available to us, for we are fortunate that nature supplies them in abundance: silk, hemp, wood, hair.

In pre-revolutionary Russia silk was used for bulletproof protection, and artificial silk is being developed today with the same purpose in mind. The plastic age, which began in earnest in the early twentieth century, has supplemented natural fibres with synthetic ones that have both advantages and shortcomings. The earliest plastics were made by trial and error; modern plastics, in contrast, are designed at the molecular level for the applications they will serve. In this chapter on the molecular aspects of materials, discussion is focussed on fibres both natural and human-made, since these provide some particularly beautiful and graphic examples of how molecular structures affect the kind of material properties that engineers worry about. But the interaction of molecular science with materials engineering is far, far broader than this.

Molecules are designed that can be converted into ultrahard ceramic materials capable of withstanding high temperatures. These are used in aerospace engineering, turbine manufacture, and power generation. Materials made from molecules (particularly polymers) are designed to conduct electricity, to capture, guide, and transform light pulses, to sieve other molecules, to protect surfaces from corrosion or contamination. They include 'smart' materials that respond to changes in their environment, which can act as autonomous switches, valves, and pumps. The impact of molecule-based materials on medicine is huge, and destined to be ever-more important: they provide artificial limbs, organs and tissues, systems for administering drugs, biodegradable sutures for surgery, sensors for monitoring the body's state of health. One day molecular engineering will make it possible to grow a new kidney or a new heart to replace a damaged one.

CABLES OF THE BODY

The idea that our bodies are a mass of cellular communities of proteins does not tally with our experience. We experience ourselves as a composite of fabrics: skin, bone, muscle, hair, fingernails. This material framework has the properties necessary to let us interact with the world. The outer layer of skin is just cladding, made up of cells that are programmed to die once they have formed the tissue. The same is true of hair, which provides insulation, and of fingernails, the evolutionary remnants of claws that once punctured and ripped. Bone and tooth are composed largely of a hard, inorganic material: calcium phosphate. Most of these materials are continuously renewed while we live; some, such as the proteins in the eye's lens, are not. These natural materials of the body serve mechanical roles and maintain our structural integrity. They are like the bricks, girders, and cladding of a building, which protect the workers from the elements and house all the complex wiring and plumbing that is necessary to conduct business as usual. Many of the body's structural fabrics are proteins. Unlike enzymes, structural proteins do not have to conduct any delicate chemistry, but must simply be (for instance) tough, or flexible, or waterproof. In principle many other materials besides proteins would suffice; and indeed, plants use cellulose (a sugar-based polymer) to make their tissues. Yet the marvel of proteins is that they are so versatile.

The molecular chains can be woven into strong fibres; cross-linked or entangled, they form the stiff matrix of horn and claw, or elastic sheets. What is more, the raw materials for making proteins are abundant in the cell. And because proteins are encoded in genes, the molecular features that give a structural protein its mechanical properties can be delicately tuned and then reliably reproduced. The most abundant structural protein in the human body, comprising about a quarter of our total protein mass, is collagen. This is a relatively simple protein whose chainlike molecules contain mostly two sorts of amino acids: glycine and proline. Glycine constitutes every third link of the chain, with proline and other amino acids (particularly lysine) in between. Some of the proline units are chemically modified, having an oxygen atom added.

This takes place in a reaction that involves vitamin C, which is why this compound is needed to maintain healthy tissues. Lack of vitamin C leads to the condition known as scurvy, caused by damaged collagen that has not been replaced. Collagen exemplifies the way in which natural protein-based structural materials differ from most synthetic polymer-based plastics. Both are composed of chain molecules; but in structural proteins these chains gather together in complex arrangements, forming thicker fibrils like ropes woven from string woven from thread. This kind of arrangement, in which structural elements are built up at successively larger scales from smaller components, is said to be *hierarchical.*

Fig. Structural hierarchy, common in nature's materials, is exemplified in the Eiffel Tower.

Engineers have learnt to use the same principle: Gustav Eiffel's iconic tower, for instance, contains struts made of a network of smaller girders, some of which are built up from even smaller girders. Collagen displays several different kinds of large-scale structure, like several different designs of tower, but all are constructed from the same basic small-scale elements. Each collagen molecular chain crimples up into a helix. Three of these twist around one another to form a ropelike triple-helical 'microfibril'. These microfibrils aggregate together in various ways. For example, they can gather in a staggered arrangement to form thick strands called banded fibrils.

The staggering creates the appearance of dark bands under the electron microscope. Banded fibrils constitute the connective tissues between cells – they are the cables that hold our flesh together. Bone consists of collagen banded fibrils sprinkled with tiny crystals of the mineral hydroxyapatite, which is basically calcium phosphate. Because of the high protein content of bone, it is flexible and resilient as well as hard. Collagen fibrils are, however, not outstandingly strong on their own, because the molecules are not linked or entangled together. But other types of collagen contain cross-linked bundles of microfibrils, joined into a kind of tough web or mesh. This provides the membrane separating the outer layers of skin from the inner layers. In contrast to the disorderly tangle of connective tissue, the eye's cornea contains collagen fibrils packed side by side in an orderly manner.

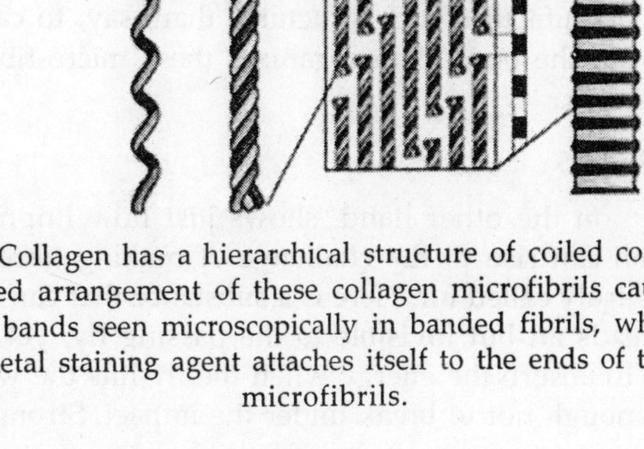

Fig. Collagen has a hierarchical structure of coiled coils. A staggered arrangement of these collagen microfibrils causes the dark bands seen microscopically in banded fibrils, where a metal staining agent attaches itself to the ends of the microfibrils.

These fibrils are too small to scatter light, and so the material is virtually transparent. The basic design principle – one that recurs often in nature – is that, by tinkering with the chemical composition and, most importantly, the hierarchical arrangement of the same basic molecules, it is possible to extract several different kinds of material properties. Collagen also constitutes the tough, elastic fabric of tendons and ligaments, and the reinforcing dentine of tooth. But the proteins on our outside – in skin, hair, and nail, as well as animal horn and hoof – are of a different kind. These tissues consist mostly of keratin, another hierarchically structured fibre. The molecular chains of keratin are again wound into helices, which are paired up in double-helical coiled strands. Two of these strands are wound together in a 'supercoil' called a protofibril, and the primary cables of keratin are composed of clusters of eight protofibrils. These fibres are surrounded by a matrix of disorderly keratin-like proteins crosslinked by sulphur atoms, like steel cables embedded in concrete. The cross-links determine the strength of the material: hair and fingernail are more highly cross-linked than skin. Curly or frizzy hair can be straightened by breaking some of these sulphur cross-links to make the hairs more pliable. Hair is a useful natural fibre; but most of the materials made from collagen and keratin proteins are formed instead into sheets (such as skin) or lumps (such as horn and hoof). They have a fibrous structure at the molecular and the microscopic scales, since it is easier for the production machinery of individual cells to manufacture such structures than, say, to cast solid blocks. But the body then organizes these micro-fibres into other shapes.

WEB DREAMS

Silk, on the other hand, shows just how impressively evolution can rise to the challenge of making fibres when they are truly called for. Here is a substance that can be spun into threads all but invisible to the passing fly, yet flexible enough to absorb the energy when the fly hits the web, and strong enough not to break under the impact. Stronger than steel or the best human-made fibres, silk is admired by the engineer for its robustness and prized by the textiles manufacturer for its exotic shimmer, its cool texture, and its ability to soak up bright dyes. The spider makes silk for many uses, and gives it a different flavour in each case. The web is spun from dragline silk; other silks are used to make supporting fibres, threads that attach the web to the branch or the rafter, strands to bind prey, strands to swaddle the developing larvae, and so forth. All of these silks are composed of protein chains in which the amino acids glycine, alanine, and serine dominate. But the precise blend of components is tailored to the silk's use. The wondrous properties of silk are a consequence of the way its protein chains are organized. Whereas in collagen and keratin the chains are coiled into helices, in silk the basic organized structural element is not a coil but a sheet. Neighbouring chains sit side by side in aligned ranks, and each chain is linked to those on either side via relatively weak hydrogen bonds, which zip the chains together into beta sheets.

Fig. Spider silk is one of the strongest known fibrous materials.

The orderly, relatively rigid sheets can stack on top of one another to create tiny, three-dimensional protein crystallites. In silk fibres these crystallites are microscopic, extending perhaps only twentymillionths of a millimetre in any direction. Beyond the crystallite regions, the protein chains continue into less orderly regions, where they are tangled together. So silk is really a kind of composite of tiny crystals dispersed in a more flexible protein matrix – a little

could do little to fine-tune these properties. Whereas delicate structural control allows nature to make several apparently different fabrics from the same raw material, chemists just got the same kind of plastic every time. In the 1950s, special catalysts were developed that provided greater control over the molecular structure of the chains, and therefore over the way that they pack.

This meant, that polyethylene could be prepared in a new, hard, high-density form with a wider range of uses than the older, softer, low-density form – as stiff drums and bottles for packaging, or as pipes and sheets in engineering and construction. These two forms differ in the degree of orderliness with which the chains are packed together; that is to say, their degree of crystallinity. As silk demonstrates, greater crystallinity results in a tougher, stiffer material, as well as making it denser and less soluble. These are desirable attributes for strong fibres. The high strength arises because molecular chains that are packed in a closer, more orderly manner cohere more avidly. So one of the principal challenges in making strong polymer fibres is to enhance the alignment of the molecules. Just about any polymer made of straight-chain that is, non-branched molecules has the potential to align its chains. But in silk this is helped by their tendency to 'zip' together to form crystalline beta sheets. In other words, the chains themselves have a kind of 'alignment instruction' programmed in. Can we make artificial polymers like this? Indeed we can.

The silk protein chains line up because of the forces of attraction between the units on different chains, which allow the chains to lock together like zips. Similar interactions exist between the chains of a class of synthetic polymers known as aramids, from which the renowned Kevlar fibres produced by DuPont are made. Kevlar is one of the best candidates so far for tethering a Space Platform. It has a tensile strength greater than that of steel, but is much lighter. The fibres are used as the strengthening cord in rubber tyres, as a fabric for bulletproof clothing, as a reinforcing material in hard composites used for aerospace engineering, and even as woven cables for anchoring oil-drilling rigs. But gram for gram, silk is stronger still. The high degree of chain alignment in a silk strand does not simply come from the fact that the molecules tend to zip together in solution. As the protein solution flows down a passageway from the spider's silk gland towards the spinneret, the polymer chains line up with the direction of liquid flow, in the same way that our hairs can be pulled into alignment by the wind.

The same process operates also during extrusion of the thread from the spinneret. It is called shearinduced alignment since the flowing liquid experiences a so-called shear force. Thus, even though it is possible to make silk-like proteins artificially, making silk threads from them is another matter entirely. If a fibre is mechanically extruded from a solution of (natural or artificial) silk protein just like pulling a thread from tacky glue, the fibre is still not as strong as real silk thread. Some researchers believe that only by building a tiny machine that mimics the spider's spinneret will we be able to make artificial silk that compares favourably with the natural material. Shear-induced alignment is used in an industrial process to make extremely strong fibres from polyethylene. The fibres are drawn from a gel-like substance in a complex procedure that results in a very high degree of alignment of the polymer chains. These fibres, sometimes called 'rocket wire', are even stronger than Kevlar, and, unlike many organic materials, they are chemically very stable. This makes them well suited for use as long-term surgical sutures.

MATERIAL GENES

One way of making artificial silk polymer is to import silk-making genes into bacteria using biotechnological techniques. Just like any other protein, silk is genetically encoded in DNA: the sequence of its amino acids along the chains is determined by a corresponding sequence of nucleotide units in a gene in the spider's chromosomes. In other words, the spider possesses a molecular blueprint for making the polymer. But notice how, because of the complexities of the spinning process, this blueprint alone is not enough to make good silk threads! The ability of living organisms to define the molecular composition of a polymer with complete accuracy is an enviable one. Modern synthetic techniques allow chemists a great deal of control over the composition of a chain – they can make hybrid polymers by grafting side chains of one chemical type onto a main chain of another, or by alternating between blocks of one type of unit and blocks of another along a single chain. But making a polymer perhaps thousands of units long in which many different units recur in a sequence of arbitrary complexity – and yet with every molecular chain in the material being identical – is something far beyond our current synthetic capabilities.

In a linguistic analogy, our state-of-the-art polymers read something like this: aaaaaaabbbbbbaaaaaaabbbbb bbaaa... Nature's polymers, meanwhile, are more like this entire sentence, and pregnant with meaning. Biotechnology allows the genes from one organism to be snipped out of the respective stretch of DNA and pasted into the DNA of another organism. The recipient then (all being well) treats the new gene as if it were its own, and uses its transcription and translation machinery to make the corresponding protein. One of the potentially valuable and, less controversial prospects of biotechnology is to transfer human genes into bacteria, which can then be bred in fermentation vats to make proteins needed for medicine. But some polymer scientists have realised that this is also a way to make protein-based materials. As well as transferring real silk genes into bacteria, one can 'write' artificial genes and express them this way. Structural proteins typically have repetitive amino-acid sequences, since the fibres they form are generally quite uniform along their length. It is relatively easy to make synthetic genes with repetitive sequences, since one need only make the individual repeat units and then join several of them together. There are ways of ensuring that the resulting synthetic gene has only a specified number of copies of the repeat unit.Using this idea, some researchers have begun to make genetically engineered protein-based materials in which the chains are programmed to adopt particular structures. 'Designed' silk-like materials have been made in this way, as well as protein-related synthetic materials similar to collagen and elastin, an elastic protein in skin.

Hybrid materials can be envisaged that marry a natural protein such as an enzyme to a synthetic protein-like chain designed to behave as a material – to pack into insoluble beta-sheet crystallites that will form a water-resistant surface coating. A possible attraction of protein-like materials for medical applications is that they would be biocompatible and biodegradable.

THE CELL'S SKELETON

Cells are laced with a network of fibres more rigid than the rope-like structural proteins collagen and keratin. These are called microtubules, and they provide the rod-like tracks on which molecular engines ferry packages around the cell. Microtubules also. There is also ongoing work to obtain silk protein from goat's milk, by transferring silk genes into goats. provide a scaffolding on which a cell alters its shape – allowing an amoeba to extend a pseudopod. The hairlike cilia appendages that protrude from cells in our respiratory tract to push dirt-filtering mucus around, and the whiplike flagella that bacteria use to propel themselves with screwlike motions through a fluid – these too are microtubules. As the name suggests, they are hollow, tubular structures. They are made up of a protein called tubulin, which is not fibrous but compact and globular. Tubulin consists of two nearly identical molecules, which pair up in a kind of dumbbell. The dumbbells stack together like bricks in a cylindrical chimney.

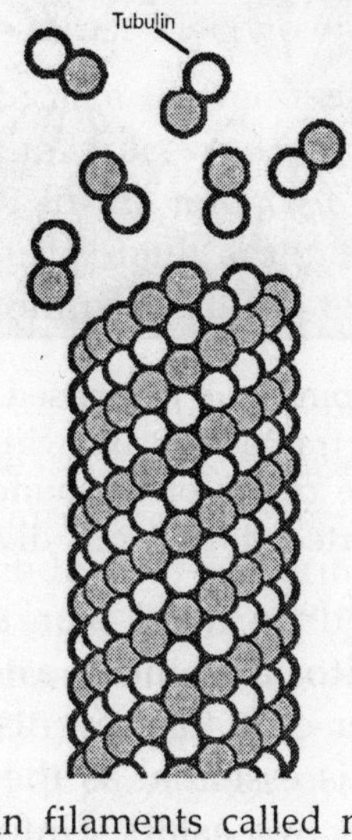

Fig. Tubular protein filaments called microtubules comprise the scaffolding of cells.

Tubulin units can attach themselves to and detach from the ends of microtubules, by which means the fibres are lengthened or shortened. An amoeba can retract its 'leg' simply by disassembling the microtubules that pushed it out. This ease of assembly enables microtubules to play a central role in cell division. Once the dividing cell has made copies of its chromosomes, it creates two clusters of radiating microtubules called asters. When the microtubules emerging from one aster meet those from the other, they merge at their ends, forming a bundle of bridging fibres between the focal points of the two asters, called the mitotic spindle cell division is called mitosis. The chromosomes are attached to this spindle and are pulled apart so that half of each is dragged to either pole. In this way the mitotic spindle provides a structure on which the duplicated genetic material can be separated out into two complete sets. The cell is then stretched and split into two halves on the framework of microtubules, each half containing a full complement of chromosomes.

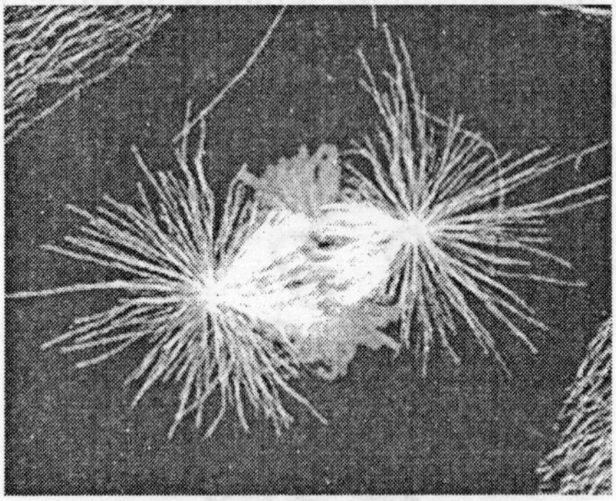

Fig. The mitotic spindle is composed of microtubules. It constitutes the framework on which chromosomes (visible here in the centre of the bundle) are arranged and sorted during cell division.

The anti-cancer drug taxol works by disrupting the assembly of the mitotic spindle and so arresting the proliferation of cancer cells. It prevents the disassembly of microtubules, which is essential as they search blindly for one another from the two asters. Unfortunately, taxol has this same effect on healthy cells; but, since cancer cells divide more rapidly, they are the hardest hit.

CARBON PIPES

Until the 1990s, little serious thought had gone into the idea of making strong synthetic fibres from tubular molecules. It is easy enough to join small molecules together in chains, but tubes look much more difficult to orchestrate. But in 1991 a Japanese microscopist named Sumio Iijima, working at the NEC Corporation in Tsukuba, discovered a kind of tubular molecule that could furnish the strongest fibres known. Called carbon nanotubes, these structures literally assemble themselves from a vapour of atomized carbon. Iijima was investigating a technique used to make carbon molecules called fullerenes, in which dozens of carbon atoms are joined together in hollow cages.

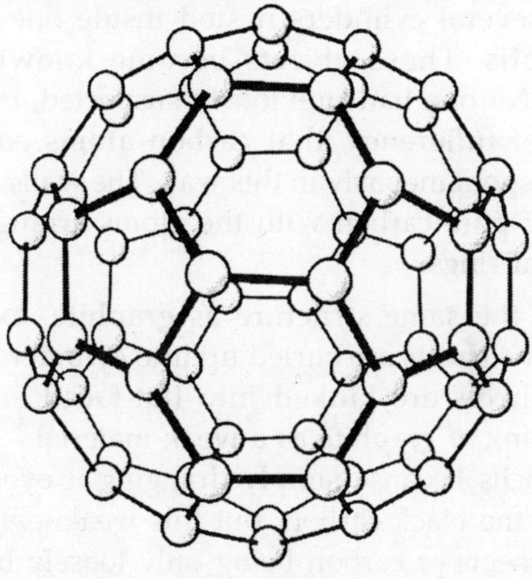

Fig. The C_{60} molecule is approximately spherical, and is composed of hexagonal and pentagonal rings of carbon atoms.

The first fullerene, a sixty-atom cage called buckminsterfullerene, was discovered in 1985. Its curious name derives from that of the US architect Richard Buckminster Fuller, who pioneered the 'geodesic dome' made from hexagonal and pentagonal facets. Buckminsterfullerene, or C_{60}, has this same structure at the molecular scale – hexagonal and pentagonal rings of six and five carbon atoms respectively, linked into a spherical cage. In 1990 a method was discovered for making fullerenes in large quantities for the first time. It involved passing an electrical discharge between two rods of graphite (which is pure carbon).

The energy of the spark vapourized some of the graphite, and the carbon atoms joined together, as the vapour cooled, to form C_{60} and other carbon cages. But when Iijima used slightly different conditions in his fullerene generator, he found something new when he examined the sooty debris under the electron microscope.

The soot was full of needle-like objects, just a few nanometres in diameter. On closer inspection, Iijima saw that these were hollow cylinders of carbon, and that each one contained several cylinders nested inside one another like Russian dolls. These objects became known as carbon nanotubes. No one had previously suspected, even after the discovery of fullerenes, that carbon atoms could arrange themselves spontaneously in this way. The walls of nanotubes are made of pure carbon with the atoms arranged in sheets of hexagonal rings.

This is the same structure as graphite, except that in nanotubes the sheets are curled up into cylinders. At the tube ends, the sheets are kinked into flat-faced end caps. We normally think of graphite as a weak material – after all, it is used in pencils because simply dragging it over paper rubs off some of the black carbon. But this weakness is the result of the flat sheets of carbon being only loosely bound to one another, so that they can slide across each other. The individual sheets, in which the atoms are tightly bound, are predicted to be extremely strong – comparable, in fact, to diamond. When these graphite-like sheets are partially joined together by chemical bonds, as they are in conventional carbon fibres, the material gains a great deal in both strength and stiffness. Carbon nanotubes are potentially the ultimate carbon fibres – the tubes are seamless sheets of graphite-like carbon.

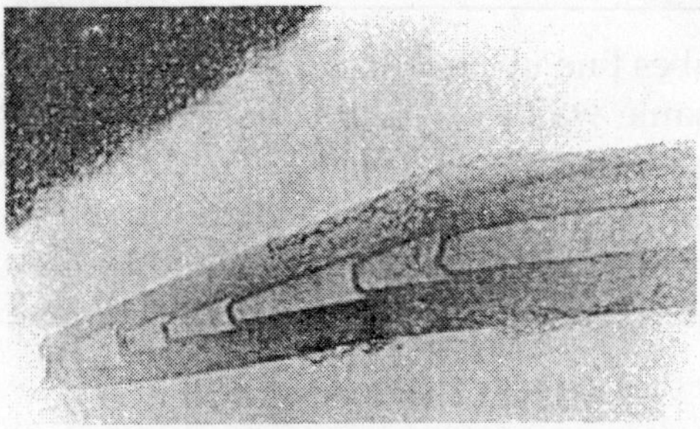

Fig. The tapering tip of a carbon nanotube, seen in cross-section in the electron microscope.

They are predicted to have a greater tensile strength than diamond; greater too than Kevlar, silk, or any other fibre natural or artificial you care to mention. Realizing this, Richard Smalley of Rice University in Houston, one of the discoverers of fullerenes, proposed that, if the Space Elevator were to be built, carbon nanotubes would hold it to the Earth. But there is a snag. So far, carbon nanotubes cannot be grown to a length of more than a fraction of a millimetre. That is not a very long tether. It also makes measurements of the true strength difficult to conduct, although some experiments at the microscopic scale have already tentatively confirmed that these tubes are indeed very strong and stiff. As strong and stiff as diamond? That is still uncertain. To make useful cables from carbon nanotubes, we will need to develop a way of controlling their growth so that the tubes can be extended indefinitely. In the synthesis of normal polymers, there is a technique called 'living polymerization' that allows the chain growth to be temporarily arrested and restarted at will, so that more and more units can be attached to ever-lengthening chains.

If someone were to develop a similar process for carbon nanotubes, it would be revolutionary. But so far there is only a hazy understanding of how the tubes grow, making it hard to see how this could be controlled. For the present, the Space Elevator will have to wait. The team made nanotube fibres and ribbons by injecting nanotubes, suspended in water by the action of soaplike molecules, into a viscous liquid polymer.

The nanotubes line up as they are injected from a capillary, in much the same way as silk proteins become aligned as they leave the spider's spinneret. The aligned tubes stick together in fibres that can be dried and handled. But these strands are not made from *continuous* nanotubes, and so they are not yet as strong as conventional carbon fibres, let alone diamond.

MOLECULES AND ENERGY

Imagine if the motor car was like the human body, so that its top speed was sustainable only for short sprints. No more driving down the motorway at let's be honest 80 miles an hour - that could only be managed for a half-mile trip to the shops. The further you had to go, the slower the speed, so as not to wear the poor vehicle out. Superficially, cars and humans are more similar than one might think, and not for the reasons Flann O'Brien would suppose. Both obtain energy by burning energy-rich fuel in oxygen; and both release an exhaust of carbon dioxide. Yet cars do not get tired when moving at close to top speed for long periods of time. Provided that you keep refilling the tank, they can keep it up almost indefinitely, whereas no sprinter could sustain their speed for an entire marathon even if they were continually chewing on glucose tablets. Why not? The question leads us into the molecular mechanisms that power the body, which are known as metabolic processes. In many ways, it is metabolism and not replication that provides the best working definition of life. Evolutionary biologists would say, If that we exist in order to reproduce - but we are not, even the most amorous of us, trying to reproduce all the time.

Yet, if we stop metabolizing, even for a minute or two, we are done for. Feel your hand. It is warm. (If it does not feel that way, try your armpit, or your tongue.) We are typically warmer than our surroundings. Whether waking or asleep, our bodies stay close to a healthy temperature of 37 °C. There is only one way of doing this: our cells are constantly pumping out heat, a by-product of metabolism. Heat is not really the point here - it is simply unavoidable, because all conversion of energy from one form to another squanders some of it this way. Our metabolic processes are primarily about making molecules. Cells cannot survive without constantly reinventing themselves: making new amino acids for proteins, new lipids for membranes, new nucleic acids so that they can divide. The wheels of the cell can never stop turning while we still live - and turning wheels consume energy. So the community of the cell is rather like the stereotypical vision of a society during the Industrial Revolution: a culture of manufacture, in which a large part of the workforce is dedicated to generating energy.

There are hundreds or even thousands of power plants in every cell of our liver, where energy production is paramount. Like the dark satanic mills of William Blake's *Jerusalem,* they produce waste as well as useful things; but cells have, on the whole, more efficient means of ensuring that they do not foul their own beds. Since energy production is essential for life, a perusal of the machinations of metabolism can be a worrisome affair. We see what a fragile thing life is - for, if only this or that process were to be disrupted, the whole system would grind to a halt, just as our social structure hangs on the premiss of an uninterrupted electricity and gas supply (not to mention clean air and sufficient water). It is a testament to evolution's inventiveness that it has shaped a machine (if you will forgive an Enlightenment metaphor) so robust that it can run for eight decades or so, God willing, before its parts begin irreparably to fail.

INTO THE FIRE

Let us begin with some simplicity. Or at least, with the appearance of the same. In the 1850s, the British scientist Michael Faraday delivered at the Royal Institution in London a series of lectures on the 'Chemical History of a Candle'. He wished to show that in the incandescent glow of the candle's flame one could read the whole of chemical science as it was then understood - and more than just chemistry. 'There is no better, there is no more open door', he said, 'by which you can enter into the study of natural philosophy'. To the trained eye, any natural phenomenon will serve as Blake's grain of sand, a window to the infinite universe. The candle will do, especially in nineteenth-century London, when showcase lectures did not have to be visually stunning.

Even combustion of paraffin wax is still not understood in all its intricate details; certainly, there are many important aspects to the problem that Faraday did not know. Yet the essence of combustion is the essence of all chemical processes that generate energy. First of all, it is a downhill process. That is the crucial point about chemical change, and about all other processes of change in the universe – they have a downhill and an uphill direction, and naturally enough they proceed in the downhill direction. What determines the topography?

In the end, it is that now semi-mystical thing called entropy. Here is the Second Law of Thermodynamics, against which no one may appeal: in all processes of change, the total entropy of the universe increases. Popular culture equates entropy with disorder. It is not a bad shorthand: a system's entropy is a measure of how many ways its atoms can be reconfigured without any noticeable difference. If someone were to come into my office and reshuffle the chaos of paper into different heaps, the chances would not be noticed. But if they were to rearrange all the papers in the carefully ordered filing system that can only be dreamed about. Crudely speaking, ordered systems have fewer indistinguishable configurations – a lower entropy – than disordered ones. The Second Law is, then, really an expression of the following: it is more likely that a system will progress from a more-ordered to a less-ordered state, simply because there are more of the latter than the former.

When one is dealing with systems that contain countless trillions of molecules, this probabilistic statement becomes a near certainty. The Second Law is a law only because its violation is overwhelmingly improbable. But of course things *do* sometimes get more ordered. Water vapour condenses into symmetrical, six-pointed snowflakes. We can go out and arrange a pile of bricks into a house, and not be arrested for transgression against cosmic law. This is true, but does not violate the principle that entropy must increase. For the Second Law applies to the universe as a whole. Order can be bought at the price of a more-than-compensating increase in chaos elsewhere. Usually this compensation is paid in heat. Yes, we can build a wall by hard manual labour – but by the time we have finished, we will have radiated enough body heat to increase the disorderly thermal motions of our surroundings sufficiently for the entropy balance sheet still to be in credit. Order is earned through disorder. Living cells maintain their organization apparently in the face of entropy's exigencies.

This led the physicist Erwin Schrödinger to suppose, in his book *What is Life?*, that the answer lay in the somewhat nebulous concept of 'negative entropy', which living organisms extract from their surroundings. This sounds suspiciously like a kind of vitalism dressed up as thermodynamics; and one still encounters today talk of enzymes as 'anti-entropy devices'. Yet there is nothing special or mysterious about the mechanics of life. Feel that hand armpit, tongue again. What are you doing, but pumping out entropy into your environment? Chemical change has a downhill direction – but sometimes it appears to go uphill. Enzymes are especially good at herding molecules uphill, although this can happen in the non-living world too. But in such cases, the reversal of direction is driven by some even more energetic downhill process. You can pull an object up a slope if it is coupled via a pulley to some more massive object moving downhill. The lighter weight rises as the heavier falls. Says chemist Peter Atkins, 'Understanding biochemistry is essentially seeking heavy weights behind subtly hiding screens.' Energy production in biochemistry is largely about pulling up as many weights as possible before a single heavy weight reaches the end of its fall. For animals like us, this heavy weight is the energy stored in the molecules we eat. Alternatively, we might draw an analogy with harnessing a river for hydroelectric power. The water is going to go on falling whatever we do – it is never going to rise spontaneously back up the mountainside. The aim, then, is to capture as much of the energy of the cascade as possible.

This is one of the major differences between our body's powerhouses and the candle's flame. In the flame, combustion rages uncontrolled, and all we get are heat and light. In the body, combustion takes place in a tightly controlled, graded sequence of steps, and some chemical energy is drawn off and stored at each stage.

BURNT SUGAR

A power station burns coal, oil, or gas – but it is much more than a flaming grate writ large. Burning is just a means to an end. The heat is used to turn water into steam; the pressure of the steam drives turbines; the turbines spin and send wire coils whirling in the arms of great magnets, which induces an electrical current in the wire. Energy is passed on, from chemical to heat to mechanical to electrical. And every plant has a barrage of regulatory and safety mechanisms. There are manual checks on pressure gauges and on the structural integrity of moving parts. Automatic sensors make the measurements. Failsafe devices avert catastrophic failure. Energy generation in the cell is every bit as complicated. A brief description cannot hope to do justice to the awesome beauty of the system.

The cell seems to have thought of everything, and has protein devices for fine-tuning it all. The main inputs are fuel and oxygen: food and air. Starve our cells of oxygen and their flame goes out. We might delicately char the fuel before we swallow it, since these subtle beginnings of combustion create compounds that delight our palate. But everything from a chocolate bar to a spinach leaf to a pig's trotter must then be broken down to a more homogeneous fuel. This happens during digestion, when enzymes in the stomach and intestines break apart our culinary creations into their raw molecular ingredients. There is a variety of energy-rich components in food, which can essentially be classified as either carbohydrates, fats, or proteins. Carbohydrates are polymers formed from molecules of the sugar glucose joined into long chains. Fats (also called lipids) may be broken down into so-called fatty acids and monoglyceride molecules such as glycerol during the digestive process. They yield twice as much energy as the same mass of carbohydrates, and the heart obtains 65 per cent of its energy from them. Glucose, then, is one of the principal 'heavy weights' of metabolism. Its descent by enzyme-assisted combustion drives the formation of energy-rich molecules called adenosine triphosphate (ATP), which are used to power other processes in the cell.

ATP is a biochemical power pack. A great many enzymatic reactions require ATP to drive them uphill. ATP is the key to the maintenance of cellular integrity and organization, and so the cell puts a great deal of effort into making as much of it as possible from each molecule of glucose that it burns. About 40 per cent of the energy released by the combustion of food is conserved in ATP molecules. ATP is rich in energy because it is like a coiled spring. It contains three phosphate groups, linked like so many train carriages. Each of these phosphate groups has a negative charge; this means that they repel one another. But because they are joined by chemical bonds, they cannot escape one another's baleful influence. Straining to get away, the phosphates pull an energetically powerful punch. The links between phosphates can be snipped in a reaction that involves water, for which reason it is called hydrolysis splitting with water. Each time a bond is hydrolysed, energy is released. Setting free the outermost phosphate converts ATP to adenosine diphosphate (ADP); cleave the second phosphate and it becomes adenosine monophosphate (AMP). Both severances release comparable amounts of energy. ATP is not the only power source for the cell's machinery, but it is the most common. Some enzymatic reactions use other, similar energy-rich molecules, particularly guanosine triphosphate (GTP). Biochemical purists will note that this is a simplified argument for why ATP hydrolysis releases energy.

GOOD DIGESTION

The business of breaking down food begins as soon as it enters the mouth, for saliva contains digestive enzymes called amylases, which start chopping up carbohydrate polymers into glucose. This is why food starts to taste sweeter when it has been chewed. In vegetables the digestible carbohydrate is largely starch; the cellulose of the plant cell walls is also a glucose polymer, but is resistant to the attack of amylases. In the stomach, the food receives more severe treatment. Here, hydrochloric acid makes the gastric juices about as corrosive as battery acid. The acid loosens up the molecular coils of proteins in the food, ready for destruction by a gastric enzyme called pepsin. The stomach is really just a storage chamber for undigested food, however. The disassembly process begins in earnest in the small intestine, into which the stomach releases its contents. The intestinal juices are spiced with a cocktail of small, hardy enzymes designed for specialized demolition jobs. While amylases attack carbohydrates, other enzymes break up the food proteins, fats, and nucleic acids. The fragments are absorbed through the intestinal lining and pass into blood and lymph vessels, which distribute these nutrients all around the body. The lining is covered in tiny folds and finger-like protrusions called microvilli, which vastly increase its surface area to ensure that nutrients are absorbed efficiently. Unfolded, the human small intestine would cover a tennis court. Digestive enzymes are produced in the pancreas, a gland with ducts that empty into the small intestine. But these potent molecules are built to break apart the very molecules from which our cells are made – so how do they not destroy our own tissues? The enzymes are manufactured with a molecular safety catch attached, which renders them inactive. In this form they are called zymogens. Not until they reach the intestine or stomach is the safety catch (a loop of polypeptide chain) removed, generally by another purpose-built enzyme.

The digestive tract of the gut is coated with a layer of mucus, which protects it from digestion enzymes. If the protective coating becomes too thin, the corrosive juices go to work on the exposed tissues, causing an ulcer. The contents of the stomach are broken down over several hours following a meal. But the body needs fuelling even after digestion is completed. So it stocks up reserves that can be called upon later. About one-tenth of the mass of our liver and 1 per cent of our muscle consists of a substance called glycogen, a compact, highly branched glucose polymer. The liver and muscle cells make glycogen from some of the sugar they receive, and store it in the form of small granules about one- to four-thousandths of a millimetre across – the pantries of the cell. If the amount of sugar in the bloodstream falls below a certain level, the body knows it is time to start feasting on its glycogen reserves. Low sugar levels trigger the formation of two hormones in the pancreas – glucagon and epinephrine – which tell cells to start breaking apart their glygocen into glucose. Too much sugar in the blood triggers a different warning system, inducing the formation of a hormone called insulin in the pancreas. Insulin is the signal for cells to start converting glucose to glycogen – to store it rather than burn it.

As an indicator of a fuel glut, insulin also signals for cells to focus on manufacturing – making proteins and fats (another emergency fuel reserve) – rather than energy generation. Insulin is a peptide (protein-like) hormone, and is genetically encoded. A relatively common defect in the genetic message leads to the production of a precursory molecule to insulin (called proinsulin) that a manufacturing enzyme cannot convert to insulin itself in the normal manner. This failure to make insulin is one of the main causes of diabetes, and means that people with the disease have to administer regular doses of the hormone to regulate their blood sugar levels.

WHEELS WITHIN WHEELS

Burning sugar is a two-stage process, beginning with its transformation to a molecule called pyruvate in a process known as glycolysis ('sugar-splitting'). This involves a sequence of ten enzyme-catalysed steps. The first five of these split glucose in half in an uphill process, powered by the consumption of ATP molecules: two of them are 'decharged' to ADP for every glucose molecule split. But the conversion of the fragments to pyruvate is a downhill affair that permits ATP to be recouped from ADP. Four ATP molecules are made this way, so that there is an overall gain of two ATP molecules per glucose molecule consumed. Thus glycolysis charges the cell's batteries. Pyruvate then normally enters the second stage of the combustion process: the citric acid cycle, which requires oxygen. But if oxygen is scarce - that is, under anaerobic conditions - a contingency plan is enacted whereby pyruvate is instead converted to the molecule lactate.

This happens, for instance, if we are exercising so vigourously that the rate of oxygen supply is unable to keep pace with the rate of glycolysis, which the body steps up to meet the high energy demand. As lactate accumulates in muscle tissues it produces the painful symptoms of a 'stitch'.

Anaerobic metabolism is a relatively inefficient way of harvesting the energy of glucose. So extreme exercise leads quite quickly to muscle fatigue: energy is consumed faster than it can be generated, regardless of how much glucose is available. It is this that limits the sprinter. The long-distance runner, meanwhile, finds a sustainable pace for which the full process of aerobic (oxygen-driven) metabolism can take its course by way of the citric acid cycle.

There is so much going on in the body that one can rarely afford a general statement, and this is no exception; for some tissues convert substantial amounts of glucose to lactate even under aerobic conditions. This process is conducted in mitochondria, sausage-shaped compartments distributed many hundred to each human cell. The first thing a mitochondrion does is convert pyruvate enzymatically to a molecule called acetyl coenzyme A (CoA). The breakdown of fatty acids and glycerides from fats also eventually generates acetyl CoA. The cycle is a sequence of eight enzyme-catalysed reactions that transform acetyl CoA first to citric acid and then to various other molecules, ending with one called oxaloacetate.

This end is a new beginning, for oxaloacetate reacts with acetyl CoA to make citric acid. In some of the steps of the cycle, carbon dioxide is generated as a by-product. It dissolves in the bloodstream and is carried off to the lungs to be exhaled. Thus in effect the carbon in the original glucose molecules is syphoned off into the end product carbon dioxide, completing the combustion process.

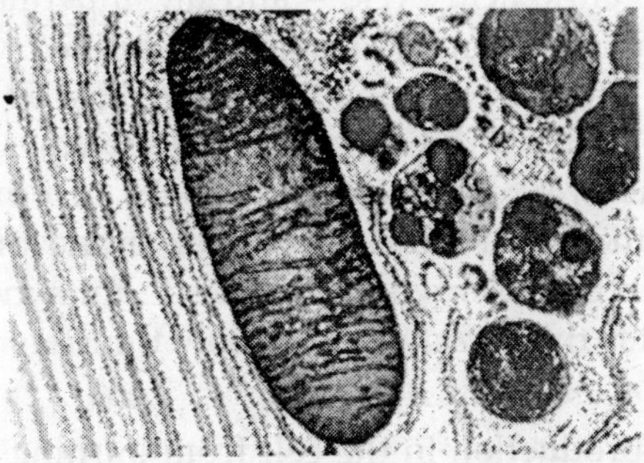

Fig. Mitochondria are self-contained compartments in cells. They are responsible for generating energy.

Animals with higher metabolic rates can sustain exertion for longer periods, because they make ATP faster. The hummingbird can flap its wings furiously almost indefinitely, like an inexhaustible sprinter. Also syphoned off from the cycle are electrons – crudely speaking, the citric acid cycle sends an electrical current to a different part of the mitochondrion. These electrons are used to convert oxygen molecules and positively charged hydrogen ions to water – an energy-releasing process. The energy is captured and used to make ATP in abundance. The electrons do not flow as if down a metal wire, however; they are carried by a molecule called nicotinamide adenine dinucleotide (NAD). Two of the reactions in the citric acid cycle add an electron and a hydrogen atom to a positively charged ion of NAD, converting it to a molecule denoted NADH. This is the electron vehicle. Transfer of electrons from NADH molecules to an oxygen molecule initiates the formation of water from oxygen and hydrogen, while regenerating NAD. This is fed back into the citric acid cycle. There are further wheels within wheels. NADH does not give up its electrons directly to oxygen. Rather, this downhill process is conducted in several stages, and each one is tapped to charge up the mitochrondrion's batteries.

The mitochrondrion is bounded by a relatively permeable outer membrane - through which come the raw ingredients that fuel the citric acid cycle - and an impermeable, highly convoluted inner membrane liberally studded with enzyme molecules. Within the inner membrane, NADH relinquishes its electrons to one of these membrane enzymes, whereupon the electrons are passed down a chain of other proteins. Eventually the electrons reach a membrane protein called cytochrome c oxidase, which has a site for binding oxygen molecules. It is this enzyme that finally transfers the electrons to oxygen.

Energy production in the cell from burning sugars takes place in two steps: glycolysis and the citric acid cycle. The first is a linear sequence of reactions; the second, taking place inside the mitochondrion, can be regarded as a series of intermeshing cyclic processes.

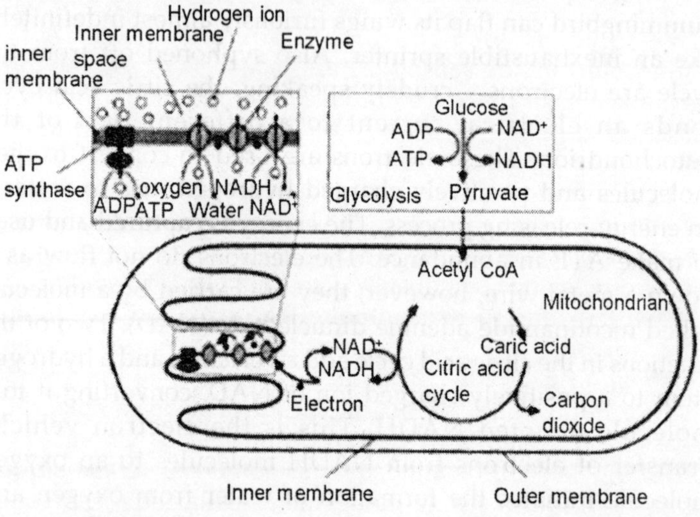

The oxygen-binding site of cytochrome c oxidase will bind certain other small molecules or ions even more strongly than oxygen - cyanide and carbon monoxide. If this happens, the electrons can no longer reach the oxygen, and so this part of the machinery ceases to turn. Jamming of this cog causes the whole of the mitochondrion's mechanism to seize up – the citric acid cycle ceases – and the system is no longer able to produce ATP. The existing stocks are exhausted in minutes. So cyanide and carbon monoxide are potent poisons, which can cause rapid death by asphyxiation. The same applies for any substances that disrupt the proteins involved in the chain of electron transport from NADH to cytochrome c oxidase. Such substances include some of the most deadly poisons; this part of the cell's energy-generating machinery is particularly sensitive to attack. The membrane proteins in the electron-transport chain pass on their bounty downhill to the next in line, and use some of the energy released to pump hydrogen ions from inside to outside the inner membrane. Thus, as the electron hops from one protein to the next, the concentration of hydrogen ions in the space between the inner and outer membrane increases.

This build-up is like the charging of a battery. The difference in hydrogen-ion concentration across the membrane, and the consequent difference in electrical charge of the two regions, is like the difference in electrical potential – the voltage drop – between a battery's terminals. Or you might think of the membrane proteins as pumps driving water back up a mountainside into a reservoir, from which energy can be extracted by letting the water run down again. The hydrogen-ion reservoir drives ATP synthesis in the mitochondrion by powering a device very much like a miniature water wheel. The hydrogen ions flow back across the inner membrane through a membrane protein called ATP synthase, which harnesses the energy to make ATP from ADP. ATP synthase has two main components. Its base, firmly lodged in the membrane, is a cylindrical channel through which hydrogen ions can flow. To the end of this channel, which opens onto the inside of the inner membrane, is attached a circular protein structure containing six globular subunits arranged in a ring.

As the hydrogen ions pass, this circular head turns, and with each revolution it makes more ATP. ATP synthase is the nub of the cell's energy-generating machinery, and the eludication of its structure and much of its mechanism of action won Paul Boyer, John Walker, and Jens Skou the Nobel Prize for chemistry in 1999. Glycolysis and the citric acid cycle are thus strikingly different kinds of metabolic process. One is anaerobic, the other aerobic. Glycolysis is potentially self-contained: one of its steps requires NAD, which is usually made in the citric acid cycle but which can be regenerated anaerobically instead by converting pyruvate to lactate. The two processes look very much like two independent kinds of metabolic pathway bolted together. This is, in all probability, precisely what they are.

The mitochondria are thought to have once been separate bacterial organisms – a notion supported by the fact that they possess their own DNA, distinct from the main genetic library in the cell's nucleus. It is believed that, around two billion years ago, the mitochrondria entered a symbiotic co-dependent relationship with single-celled organisms that metabolized anaerobically using the glycolytic pathway. Around this time, the spread of green algae primitive plant life led to a sharp increase in the amount of oxygen in the Earth's atmosphere; before then, there was rather little oxygen in the air.

The mitochondria-like bacteria were able to 'breathe' oxygen using the citric acid cycle. Symbiosis between the anaerobic cells and the aerobes arose because the anaerobic cells made pyruvate, which the aerobes used for fuel, while the aerobes produced NAD to help drive the anaerobes' glycolytic process. Eventually, the anaerobes engulfed the aerobic organisms to become single, composite oxygen-breathing cells. The rest, as they say, is history. Yeast never took this step: it is still a glycolytic anaerobe, living off sugar. But instead of turning pyruvate into lactate, yeast cells convert it to ethanol, to the delight of brewers for millennia. This is the process of fermentation, from which stemmed the whole of our understanding of enzyme action. It too generates carbon dioxide, the end product of combustion, which bubbles out of the brewing vats of the world.

SOMETHING IN THE AIR

Oxygen enters the body through the lungs. But it is not very soluble in blood, and so cannot be carried to the mitochondria simply by dissolving. (Carbon dioxide, in contrast, is soluble enough to make its own way from cells out to the lungs in the bloodstream.) Oxygen is carried by red blood cells packed with the protein molecule haemoglobin, which has a strong avidity for oxygen. Red blood cells could hardly be more single-mindedly designed for their job. Unlike other cells, they contain no DNA or RNA, no compartments and almost no other enzymes – they are basically sacks of haemoglobin. All the genetic and enzymatic machinery required for making this protein is ejected at the last moment as the cells mature. The red colour of blood is akin to the red of rust, the hallmark of iron.

At the heart of the haemoglobin molecule are iron atoms, caught in washer-shaped molecular traps called porphyrins. These iron-loaded porphyrins are known as haeme groups; they absorb blue-green light strongly and so look bright red. Each haemoglobin molecule contains four haeme groups, whose iron atoms provide attachment sites for oxygen molecules. As an oxygen transporter, haemoglobin must be able to bind its cargo tightly, but also to let it go again. How can it do both? Haemoglobin employs a clever trick that is often used when a protein has to modify its behaviour once it has bound its target molecule. Put simply, one part of the protein can 'feel' what is going on in another part. In the blood vessels of the lung, where oxygen is plentiful, the attachment of one oxygen molecule to a haeme group sends a shudder through the protein that encourages the other three haeme groups to take up oxygen too. Haemoglobin gives up its oxygen primarily in muscle tissue, where it transfers the molecule to myoglobin, another oxygen-binding protein, which has an even higher affinity for oxygen. Once one oxygen molecule has been taken from haemoglobin, a 'reverse shudder' weakens the grip of the other three haeme groups on their own cargo, and they release the oxygen more readily. These shudders – small but significant changes in the folded shape of the protein chain – are called allosteric motions. All vertebrate animals and many invertebrates transport oxygen using haemoglobin, although the exact shape of the protein differs in different species. Arthropods and molluscs use a different oxygen-binding protein called haemerythrin, which also contains iron but not bound in a haeme group.

This has a violet-pink colour when charged with oxygen, but is colourless when not. Some marine invertebrate creatures use a different metal for transporting oxygen: their oxygen-binding protein, called haemocyanin, is blue, betraying the presence of copper. They are the true blue-bloods of the ocean.

LEAF POWER

The oxygen-rich atmosphere of Earth is the work of plants. It is a by-product of photosynthesis, the biological use of the sun's energy to make molecules. Since plants and photosynthetic bacteria lie at the base of the food pyramid, all life on Earth is ultimately solarpowered. Without plants we are done for: we are given not our daily bread, and our cattle starve in barren pastures. Photosynthesis is an old, old process. Algae were photosynthesizing at least three and a half billion years ago, when the continents were newly formed and ungreened. These organisms were the first *autotrophs:* 'self-feeders', making their own molecules from little more than light, water, and carbon in the air. About 60 billion tons of carbon are plucked every year from the carbon dioxide in the atmosphere and are turned into energy-rich biomass.

Some of this we eat, burn, arrange into houses and tables, feed to livestock, pulp into paper, spin into cloth. Much of it falls, is broken down by micro-organisms, and is released back into the atmosphere as volatile carbon compounds. Over geological time, some will be buried, compressed into coal, or decomposed into oil or gas. Photosynthesis depends on molecules that interact with light, absorbing some of its energy and channelling it into chemical processes. While mammalian cells have fuel-burning factories in the form of mitochondria, the solar-power centres in the cells of plant leaves are compartments called chloroplasts.

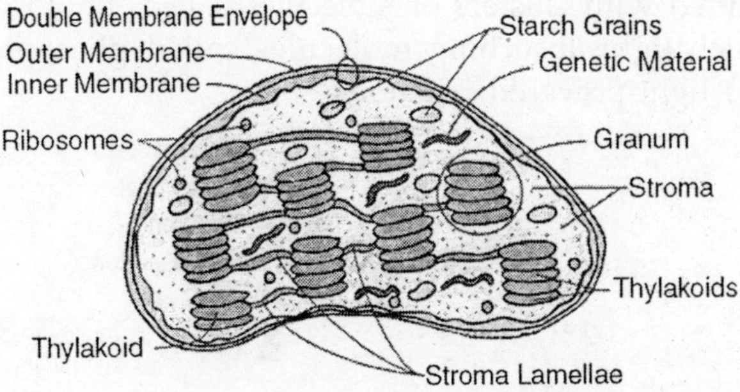

Fig. Plants capture sunlight and convert it to chemical energy in chloroplasts, filled with the stacked sheets of thylakoid membranes.

The process conducted herein is, broadly speaking, the reverse of glucose metabolism. The chloroplast takes carbon dioxide and water, and from them constructs the sugar. 'Burning' glucose is an energetically downhill process, so it follows that the manufacture of glucose in photosynthesis is uphill. This is why the plant needs the energy of light rays to do it. Yet the plant uses this energy not just to create glucose for weaving into the cellulose walls of its cells, but also - and just as importantly - for making ATP molecules to drive the cells' chemistry. There are several similarities between the processes of aerobic metabolism and photosynthesis. Both consist of two distinct subprocesses with separate evolutionary origins: a linear sequence of reactions coupled to a cyclic sequence that regenerates the molecules they both need. The bridge between glycolysis and the citric acid cycle is the electron-ferrying NAD molecule; the two sub-processes of photosynthesis are bridged by the cycling of an almost identical molecule, NAD phosphate (NADP).

In the first part of photosynthesis, light is used to convert NADP to an electron carrier (NADPH) and to transform ADP to ATP. This is effectively a charging-up process that primes the chloroplast for glucose synthesis. In the second part, ATP and NADPH are used to turn carbon dioxide into sugar, in a cyclic sequence of steps called the Calvin–Benson cyclez'. The first process takes place at the surface of a folded membrane inside the chloroplast, **called** the thylakoid membrane. This is studded with clusters of molecules called 'photosystems', in which light-absorbing molecules called photopigments initiate light-powered reactions.

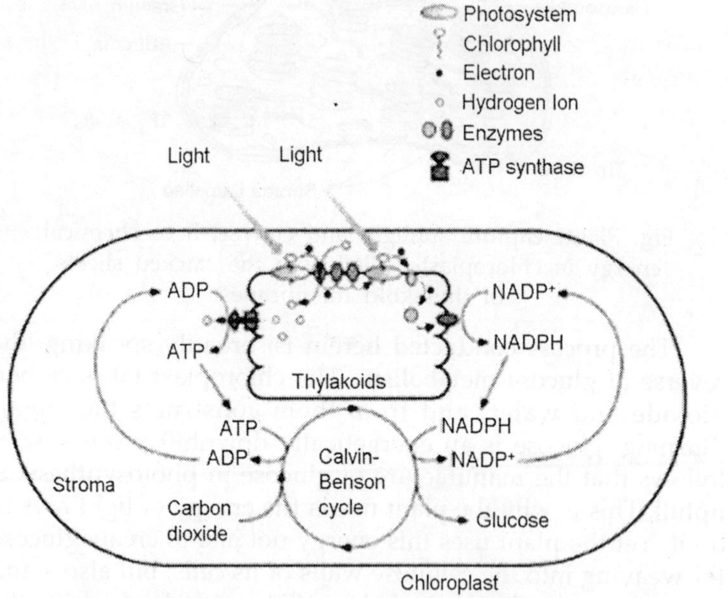

Fig. During photosynthesis, captured solar energy is used to split water and to make ATP, which then drives the conversion of carbon dioxide to glucose

At the heart of the photosystems, called the photosynthetic reaction centre, is a molecule called chlorophyll *a*. This absorbs red and blue light strongly, and is thus responsible for the green colour of leaves. When chlorophyll receives light energy, it becomes 'excited', like an apple tree that gets shaken. In its excited state, chlorophyll is less able to hold on to its outer electrons, and one of them comes free. The electron is passed on to an enzyme; once the enzyme has gained two electrons from 'shaken' chlorophylls, it can transform a positively charged ion of NADP to NADPH.

The electron-deficient chlorophylls are then replenished with electrons plucked from water molecules in another light-powered reaction. The water is broken into hydrogen ions and oxygen atoms. The oxygen atoms combine into two-atom oxygen molecules, which the plant releases through openings in the leaf surface. The electron taken from water is passed to chlorophyll along a chain of molecules embedded in the thylakoid membrane. Each transfer step is a downhill process that releases energy, some of which is tapped to pump hydrogen ions into the inner space of the thylakoid membrane. This imbalance is then harnessed as an energy source by ATP synthase molecules lodged in the membrane, which perform their windmilling conversion of ADP to ATP. But the task of photosynthesis is not finished with water-splitting and the production of the energy source ATP and the electron source NADPH. These two ingredients are released into the fluid outside the thylakoid membrane, called the stroma, where they drive the reactions of the Calvin–Benson cycle, which turn carbon dioxide to sugar.

These processes are called the 'dark reactions', because they do not require light directly. In 1961 US chemist Melvin Calvin won the Nobel Prize for deducing most of this sequence. Chemists are currently interested in designing artificial molecular systems that, like chloroplasts, harness sunlight to drive chemical synthesis. A team at Arizona State University has mimicked the chloroplast in synthetic cell-like (and cell-sized) structures called liposomes: hollow, spherical membranes made from lipid molecules. The researchers peppered liposome membranes with designed molecular assemblies that perform the same task as photosynthetic reaction centres - using light energy to pump hydrogen ions into the liposome's hollow interior. The researchers inserted molecules of ATP synthase into their liposomes, which release the hydrogen ions and make ATP into the bargain. The hope is that the energy stored in ATP can be tapped for chemical synthesis - to conduct some industrially useful biochemical reaction.

ENDING WITH A BANG

Glucose and candle wax a hydrocarbon are both 'embodied energy': breaking their bonds with oxygen releases energy as heat unless channelled into other forms. But some chemists seek molecules that pack a bigger punch. How much energy can be crammed into a molecule? Alfred Nobel occupied himself with that question in the nineteenth century, and the resulting irony - the fortune amassed from the invention of dynamite, which now funds an annual peace prize - is well known. Nobel's innovation was the invention not of an energyrich molecule but of a means for packaging an existing explosive into a form that was less likely to blow up in one's face. The oldest explosive is gunpowder, a mixture of sulphur, nitre potassium nitrate, and charcoal devised in China around the eleventh century AD.

This was deployed in the West with little change and terrible effect until the nineteenth century, when scientists began to uncover ways of making a bigger bang. In 1845 the Swiss chemist Christian Schönbein discovered nitrocellulose, a compound made by treating cotton fibres (cellulose) with nitric and sulphuric acid. This was the first 'semi-synthetic' polymer, a product both of nature and of the chemist's art. It was later developed into celluloid, a hard plastic, and rayon, the first 'artificial silk'. But nitrocellulose was also explosive, and became known as gun cotton. Gun cotton's violent nature was hard to control; early attempts to manufacture it led to several deaths. In 1847 the Italian chemist Ascanio Sobrero synthesized a similarly hazardous substance called nitroglycerine. Alfred Nobel began to study this compound in 1859, attempting to find a way of rendering it stable until deliberately detonated. Persisting despite an explosion in 1864 that killed his younger brother, Nobel found that mixing nitroglycerine with a clay called kieselguhr produced a putty-like explosive that could be handled safely.

He called it dynamite. In 1875 Nobel introduced gelignite or 'blasting gelatine', a jelly-like mixture of nitroglycerine and gun cotton, which was more powerful than either substance on its own. These explosives were used largely for mining and construction blasting, and they made Nobel's fortune. But inevitably they were adopted for military use too. The British and French armies of the late nineteenth century used cordite, an explosive similar to blasting gelatine; the German troops took to trinitrotoluene (TNT), which explodes only if detonated with a secondary explosive. In *Brave New World* Aldous Huxley tells us of the rare chemistry of this lethal combination:

$CH_3C_6H_2(NO_2)_3 + Hg(CNO)_2$ = well

An enormous hole in the ground, a pile of masonry, some bits of flesh and mucus, a foot, with the boot still on it, flying through the air and landing, flop, in the middle of the geraniums.

All of these compounds are organic substances containing the nitro group: a nitrogen atom with two oxygen atoms attached. The explosiveness of nitro compounds derives from the fact that nitrogen atoms recombine to form nitrogen molecules when the materials are ignited. These molecules have very stable bonds whose formation releases a lot of energy. The oxygen atoms, meanwhile, stimulate the combustion process, allowing it to happen very quickly. Developing more powerful explosives has been largely a matter of finding ways to pack more nitro groups into an organic compound. RDX or cyclonite, an explosive used in today's weaponry, improves on TNT in this respect. The most powerful explosive in current production is a nitrogen-rich compound called HMX, an abbreviation for 'high-melting explosive'. In early 2000, a potentially still more energy-packed nitro compound was devised by chemists at the University of Chicago. Called octanitrocubane, it consists of a cube of eight carbon atoms, to each of which is attached a nitro group. Not only is this molecule extremely nitro-rich, but the cubic shape means that the bonds between carbon atoms are highly strained and easy to burst open. And the compact molecules should be able to stack together into a very densely packed crystalline form.

The initial experiments have not yielded this high-density form, but calculations predict that, if it can be made, it will have an explosive energy content greater, gram for gram, than any known non-nuclear explosive. Alfred Nobel's prize-giving bequest seems to have been a gesture provoked by the regret he felt for the lethal and destructive applications of his work. Clearly, not all chemists will feel this way about weapons-related research. My own view, for what it is worth, is that such work is arguably a form of scientific misconduct. But, more importantly, work on high explosives demonstrates that scientific research does not split cleanly into an amoral 'pure' science and a dirty and socially accountable 'applied' science or technology. Making octanitrocubane was a feat of technical brilliance; it was also funded by the US Defence Department. Explosives reveal the Janus face of molecular science.

3

Molecular Communication

Each of us is a new world. The molecular view of life reveals that the appropriate analogy for a cell is a city, teeming with molecular inhabitants. Our many-celled bodies are thus collaborations between communities. One cell communicates with another as London does with Liverpool, New York with Philadelphia: messages are passed down wires or carried from place to place by courier. Goods are transported hither and thither along the transportation network of the blood and lymph circulatory systems. It is just as Berzelius said: 'This *power to live* is the result of the mutual operation of the instruments and rudiments on one another.' Molecular biology has long been content merely to document the social webs of the cell: deducing which molecules speak to which, and how they come and go. But ultimately this is not enough. We need to know also what is said, and how the messages are passed on one to another. This is information that a pharmaceutical chemist can use to develop drugs.

The fundamental challenge for medical science is to learn how to take part in the body's molecular conversations: to intercept harmful or unpleasant messages, to send out new warning messages, to prevent undesirable interactions. It is largely as a result of these efforts in the field of biochemistry that chemistry itself is undergoing something of a reinvention. The imagination of chemists is fired up by what they see to be possible in biology. Although much of the chemicals industry is devoted to the manufacture of 'passive' products

- new plastics, cement, glue, paint, synthetic fibres - drug molecules were always a little different. For their task is to partake in a dynamic process, to engage in the active life of the cell. They are like actors primed for a dramatic role; indeed, they often play their part by impersonation. Now chemists are beginning to appreciate that this kind of dynamism can be achieved in purely synthetic chemical systems too.

And so chemistry is becoming less about the properties of individual molecules and more about how groups of different molecules behave together - forging and breaking relationships, modifying each other's tendencies, sending out signals. Chemistry is becoming a science of *process*.This is the attitude that underpins much of the work is discussed in the book: the development of molecular solar cells, of chemical sensors, of a molecular nanotechnology, and of molecular devices that process information. Much of the research in this area is gathered under the umbrella of *supramolecular chemistry*, which means chemistry beyond the molecule: the science of molecules in communication.

The ways in which molecules communicate in biology, before providing a glimpse of how a similar gregariousness can be encouraged in synthetic molecules. As ever, we must remember that, while nature is inspirational, it is also parsimonious and blind. Biology uses a limited range of materials, and has a tendency to adapt a good solution endlessly for new purposes rather than exploring a completely new avenue each time. Just as the Jumbo jet is not a scaled-up pigeon, so the wise molecular engineer takes from nature principles but not blueprints.

MOLECULAR MAIL

Italy and Germany became countries when their patchworks of small kingdoms and city states agreed to respect a central authority - to collaborate towards a greater good for all. The body cannot behave as a coherent entity unless cells do likewise. This means that there must be mechanisms for sending commands, edicts, and calls to action throughout the entire realm. Nerve signals from the brain are one means by which the body coordinates its actions. They are the body's telephone system. But general messages sent *globally* throughout the body are posted like a mass mail shot into the bloodstream, in the form of molecules called *hormones*. These are diverse both in form and in function. Some hormones are large proteins; others are small organic molecules. Some are soluble in water, others insoluble which means that courier molecules are required to carry them through the bloodstream.

Some convey urgent, immediate messages such as 'Run away!' Others have long-term effects, promoting growth or the development of sexual characteristics. All hormones are the product of the *endocrine system*, a series of glands that constitutes the overarching regulatory system for the whole body. We have seen already how the hormones insulin and glucagon, produced in the pancreas, control the blood's sugar content. Similarly, the rate of metabolic processes in cells is regulated by the hormones thyroxine and triiodothyronine released from the thyroid gland. These hormones affect energy production and oxygen consumption, in part by altering the heart-beat rate. The control centre of the endocrine system is the hypothalamus, a gland in the brain. The hypothalamus is connected to the pituitary gland, which sits just below it, from where hormones are dispatched to other glands.

For example, a fall in metabolic rate prompts the hypothalamus to send a molecule called thryotropin-releasing hormone to the pituitary. This gland in turn starts to send out thyroid-stimulating hormone, which triggers the thyroid gland into action. All of the hormones released from the pituitary are peptides: small protein-like molecules. Antidiuretic hormone, for instance, controls the body's water content by regulating the production of urine in the kidneys. Growth hormone provides a stimulus for cell multiplication, and plays a key role during childhood and adolescence. It also stimulates localized growth of tissue when repairs need to be made – during wound healing.

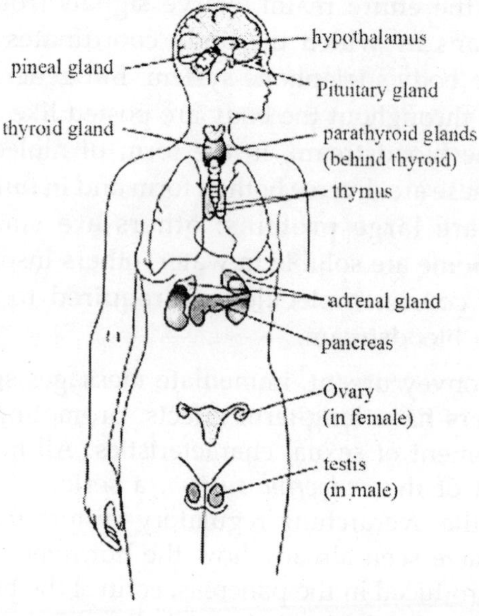

Fig. The body's endocrine system, a series of hormone-producing glands.

The adrenal glands manufacture some important steroid hormones. These are insoluble molecules with a carbon backbone consisting of several small rings joined together. Some steroids, such as cortisol, regulate the storage and use of the body's energy resources: the conversion of glucose to glycogen and the breakdown of proteins into amino acids. Bodybuilders and athletes use these hormones legally or otherwise to build up body mass and muscle. As you might anticipate, the hormone adrenaline is also a product of the adrenal glands.

Along with noradrenaline, it is released rapidly into the bloodstream in response to stress. Both hormones quicken the heart rate and dilate the blood vessels, increasing the oxygen supply to muscles so that they are prepared for extreme exertion. The sex glands - ovaries in women, testes in men - release the hormones that differentiate the sexes and trigger changes in growth during puberty. Testosterone stimulates sperm production in men. Oestrogen and progesterone control the female menstrual cycle; their production is regulated by hormones released from the pituitary gland, called follicle stimulating hormone (FSH) and luteinizing hormone (LH). These two hormones regulate ovulation during the menstrual cycle. In the early days of pregnancy, high levels of oestrogen and progesterone in the bloodstream inhibit the production of FSH and LH and suppress ovulation.

Birth-control pills have the same effect: containing oestrogen and progesterone, they persuade the woman's body that she is already pregnant. Production of oestrogen declines when a woman is in her thirties, and particularly during the menopause. Side effects of low oestrogen levels include an increased susceptibility to coronary heart disease and to bone loss, which are two of the prime motivations for administering oestrogen in hormone replacement therapy. The treatment remains controversial, because long-term doses of oestrogen can have unwelcome side effects of their own, including an enhanced susceptibility to breast cancer and different forms of heart disease.

SWITCHED ON

How is a hormonal message read? This depends on the nature of the message. Some hormones can be posted right through the walls of cells, wherein they bind to some receptor protein. This activates the receptor to stimulate the transcription of a particular gene, making a protein that the cell needs. This so-called direct gene mechanism of hormone action works for hormones that are small and insoluble, so that they can penetrate the fatty cell membrane. But many hormones, particularly those comprised of peptide and protein molecules, get no further than knocking on the doors of the cell. They are received by butlers at the cell surface – receptor proteins whose job it is to convey the message to others inside the cell. Like most other molecular communication, the passing of a message from hormone to receptor protein is an intimate affair. Molecules show no inhibitions: they speak to one another through close embraces.

Lacking any other means of recognition, molecules identify one another by 'touch', through binding events in which the receptor latches onto a target (substrate) with precisely the right shape, like a key fitting in a lock. Each hormone-receptor protein on a cell surface has a binding site sculpted to fit around the hormone. Despite the variety of messages that hormones convey, the mechanism by which the signal is passed from a receptor protein at the cell surface to the cell's interior is the same in almost all cases. It involves a sequence of molecular interactions in which molecules transform one another down a relay chain. In cell biology this is called *signal transduction.* At the same time as relaying the message, these interactions amplify the signal so that the docking of a single hormone molecule to a receptor creates a big response inside the cell. It works like this. The receptor proteins span the entire width of the membrane; the hormone-binding site protrudes on the outer surface, while the base of the receptor emerges from the inner surface.

When the receptor binds its target hormone, a shape change is transmitted to the lower face of the protein, which enables it to act as an enzyme. The process that the enzyme catalyses is the 'activation' of a so-called G protein, attached to the inner surface of the membrane. G protein is short for guanine-nucleotide-binding protein: the protein holds onto a molecule of guanosine diphosphate (GDP). When a hormone-charged receptor interacts with a GDP-laden G protein, the G protein first replaces the GDP with guanosine triphosphate (GTP, analogous to energy-rich ATP), and then breaks in two. The half that binds the GTP becomes an enzyme, and travels off to activate another enzyme at the inner surface of the cell wall. Commonly, this other is adenylate cyclase, a protein that converts ATP into cyclic AMP (cAMP). The participants of all these processes are stuck to the cell wall. But cAMP floats freely in the cell's cytoplasm, and is able to carry the signal into the cell interior. It is called a 'second messenger', since it is the agent that relays the signal of the 'first messenger' (the hormone) into the community of the cell.

Cyclic AMP becomes attached to protein molecules called protein kinases, whereupon they in turn become activated as enzymes. Most protein kinases switch other enzymes on and off by attaching phosphate groups to them – a reaction called phosphorylation. The action of a protein kinase initiates a cascade of reactions, since each activated kinase can act on several enzyme molecules, each of which in turn can do its job many times. In this way docking of one hormone to its receptor can affect many molecules inside the cell: the signal becomes amplified.

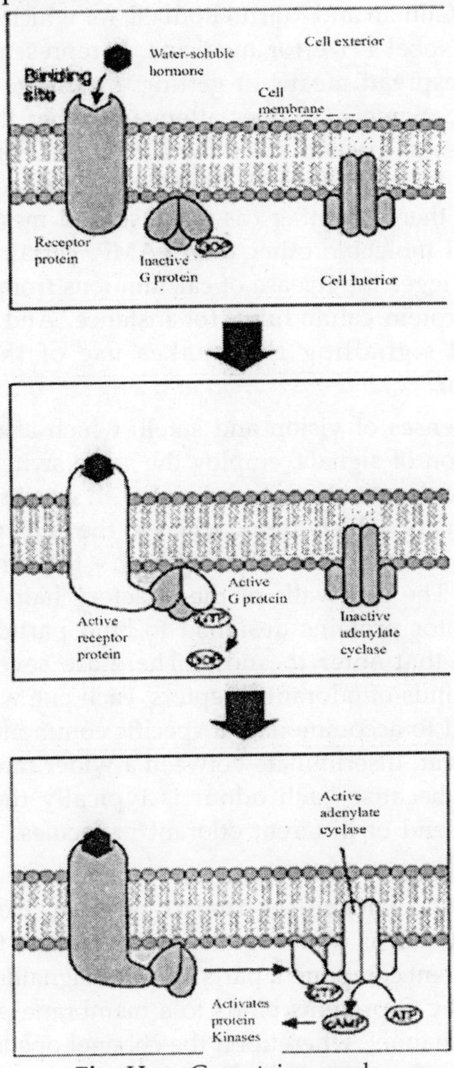

Fig. How G proteins work.

The process might sound rather complicated, but it is really nothing more than a molecular relay. The signal is passed from the hormone to its receptor, then to the G protein, on to an enzyme and thence to the second messenger, and further on to a protein kinase, and so forth. The G-protein mechanism of signal transduction was discovered in the 1970s by Alfred Gilman and Martin Rodbell, for which they received the 1994 Nobel Prize for medicine. It represents one of the most widespread means of getting a message across a cell membrane. Some hormones attenuate rather than stimulate cell processes; in such cases the activated G proteins might exert an inhibitory effect on their target enzymes rather than activating them. In other cases the second messenger might be a small molecule other than cAMP: certain activated G proteins trigger the release of calcium ions from the calcium-binding protein calmodulin, for instance. And it is not just hormonal signalling that makes use of the G-protein mechanism.

Our senses of vision and smell, which also involve the transmission of signals, employ the same switching process. The roof of the nasal cavity is lined with smell sensors called olfactory hairs, which are attached to the ends of nerve cells that carry signals to the olfactory bulb – the 'smell centre' of the brain. The cell walls of the olfactory hairs are studded with receptor proteins designed to bind particular odorant molecules that enter the nose. There are several hundred different kinds of odorant receptors, each one with a binding site shaped to accommodate a specific common odorant. We can, however, discriminate between a wider range of odours than this, because each odour is typically the result of a complex blend of different odorant molecules.

The olfactory bulb forms an 'image' of the smell from the mixture of impulses that it receives from different receptors, much as we recognize a person's face from the sum of the different component parts. In smell signalling, the cAMP produced by G proteins binds to a membrane protein called a sodium channel, whereupon the channel opens up and lets sodium ions flow into the cell. This triggers a nerve impulse, which passes to the olfactory bulb. The same basic process generates visual signals in the optic nerve when stimulated by light. Our sense of taste is due largely to our olfactory system. The taste buds in our tongues can distinguish only relatively crude signifiers of taste: sweetness, bitterness, saltiness, and sourness. The full delight of a matured cheese or freshly baked bread comes mostly from the odorant molecules they release.

ALL IN THE MIND

Hormones can trigger complex webs of biochemical action, but the messages they bear are pretty crude, related to the exigencies of growth and survival. It is quite another matter that communication between molecules gave birth to the Sistine Chapel, to *The Magic Flute,* to the theory of relativity. Yet the mind is, after all, made of molecules. At the same time, the mind is still a mystery – one of the great remaining mysteries of science. Some scientists argue that the mind will never be able fully to comprehend itself, that the self-referential nature of the problem will always create blind spots.

Others believe that a scientific explanation of consciousness is on the horizon. Either way, it is likely that the secrets of the mind lie far beyond the molecular realm, embedded in questions about the behaviour of complex, highly connected information networks. Here we see the limitations of reductionism – for the molecular processes of thought are now fairly well mapped out, yet their collective consequences are barely sketched. The brain contains somewhere in the region of a billion to a hundred billion brain cells or *neurons.* That is nothing to write home about – other organs are comparably populous. But the distinguishing feature of the brain is the complexity of the communication network between these cells.

Each neuron makes around a thousand links, so there may be up to a hundred trillion interconnections in the brain – about the same as the number of stars in a thousand galaxies like our own. On such a transportation network, you would be lost in an instant. The degree of connectivity in the integrated circuits of computers is nowhere near so great, and it is no surprise that computers, for all their literal-minded speed, fail miserably at some tasks that a child can do in a flash. Neurons send nerve signals - in essence, electrical pulses - to one another along tubular channels called axons.

The axon ends in a series of branches whose tips push up against the membranes of other neurons. At these junctions, called synapses, a nerve signal is transmitted from one neuron to another. Neurons also sprout many shorter, bushy branches called dendrites, which collect information from the axons of other cells. The axons are, if you like, the motorways of the brain, stretching from one neuronal city to the next. They end in slip roads that connect up at synapses to the city road system of the dendrites. Although axon signals are electrical, they differ from those in the metal wires of electronic circuitry.

The axon is basically a tubular cell membrane decorated along its length with channels that let sodium and potassium ions in and out. Some of these ion channels are permanently open; others are 'gated', opening or closing in response to electrical signals. And some are not really channels at all but pumps, which actively transport sodium ions out of the cell and potassium ions in.

These sodium-potassium pumps can move ions 'uphill' - from regions of low to high concentration - because they are powered by ATP. In its 'resting' state, the axon has an imbalance of sodium and potassium ions inside and outside that sets up a charge difference, or voltage, across the membrane: the fluid inside has a small negative charge (the 'resting potential') relative to that outside. When a signal is sent down the axon, some of the gated sodium channels open up, altering the distribution of ions and reversing the imbalance: the inside becomes positively charged relative to the outside.

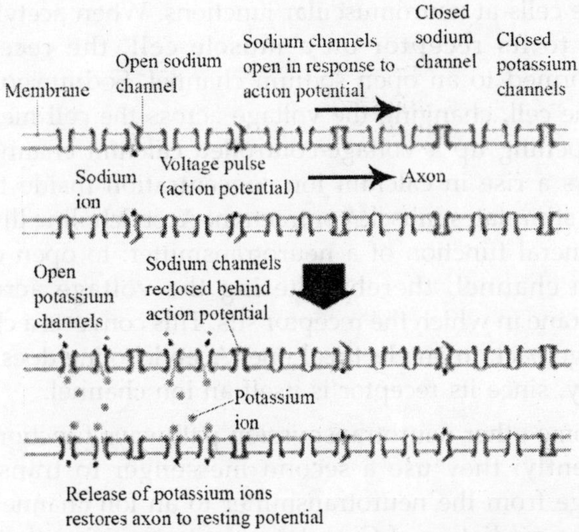

Fig. Electrical pulses are sent down the axon by the opening and closing of ion channels.

This region of reversed voltage opens up sodium channels ahead of it, so that it moves along the axon. At the same time, the channels behind close up and the resting potential is restored. In this way, a voltage pulse or 'action potential' travels down the axon. At a synapse, this nerve impulse is transmitted from the axon to another neuron. The signal is generally first converted from electrical to chemical form. A small molecular messenger called a neurotransmitter conveys the signal across the space called the synaptic cleft between the terminal membrane of the axon and the membrane of the other neuron. The neurotransmitter is packaged up inside a bubble-like membrane that merges with the axon's cell wall, releasing the molecular message into the synaptic cleft. It travels to the outer surface of the other neuronal membrane, where it becomes bound to receptor proteins. A diverse array of molecules serve as neurotransmitters. Some are simple amino acids – glycine and glutamate – or molecules derived from them, such as serotonin and dopamine.

The molecule acetylcholine is a neurotransmitter that carries messages between the central nervous system and muscle cells at neuromuscular junctions. When acetylcholine binds to its receptor on a muscle cell, the receptor is transformed to an open sodium channel. Sodium ions rush into the cell, changing the voltage across the cell membrane and opening up a voltage-controlled calcium channel. This triggers a rise in calcium ion concentration inside the cell, which stimulates muscle contraction. Acetylcholine illustrates the general function of a neurotransmitter: to open or close an ion channel, thereby altering the voltage across the membrane in which the receptor sits. This converts a chemical message back to an electrical one. Acetylcholine does the job directly, since its receptor is itself an ion channel.

Some other neurotransmission pathways function rather differently: they use a second messenger to transfer the message from the neurotransmitter to an ion channel, again with the mediation of G proteins. Is it surprising that the G-protein signal transduction mechanism appears in so many different contexts? Not really. As the complexity of multicelled organisms evolved, cells with ever more specialized functions arose from common ancestors with more general functions. Tried-and-tested mechanisms for certain tasks would be retained, though adapted where necessary. That, after all, is why we share genes with yeast and bacteria. The G-protein pathway is an effective way of passing a chemical message from the outside to the inside of a membrane, and amplifying it in the process. The cell's motto is: if it works, find a way to use it.

PLEASURE AND PAIN

Neurotransmission is a common target for drugs - beneficial, harmful, pleasurable, or, depending on the dose, all three. The nervous system is one of the most vulnerable parts of the body: if nerve impulses are blocked, we cannot move. Many animals make toxins that cause paralysis in their prey by attacking the sodium-potassium pumps or the voltage-gated ion channels in axons, blocking the progress of action potentials. Muscle action is also affected by drug molecules that resemble acetylcholine and so compete with it in binding to the receptor proteins at neuromuscular junctions. Nicotine, the active ingredient of tobacco, is one such: it binds to a certain class of acetylcholine receptors in muscle and causes the associated stimulatory sensations: increase in heart rate and dilated pupils. Why the sensation is pleasurable is not, however, fully understood.

Curare is a lethal toxin present in the bark of a South American plant, which was once extracted and used by the indigenous people to poison arrow tips. Curare binds to the same class of acetylcholine receptors as nicotine, but does not activate them - so muscle action is prevented. An animal poisoned with curare will die of asphyxiation, unable to inflate its lungs. The medieval poison hemlock works in the same manner. Whereas some neurotransmitters stimulate neurons, the role of others is to quieten them: to suppress the firing of action potentials. These are said to be inhibitory, and include glycine and the molecule gamma-aminobutyric acid (GABA). Our thoughts are a complex interplay of stimulation and inhibition, as neurons weigh up the various signals they receive from their neighbours and decide whether or not, on balance, they should fire off a salvo themselves. Hallucinogenic drugs such as LSD (lysergic acid diethylamide) and mescaline overexcite the brain by enhancing the stimulatory effects of serotonin.

The poison strychnine blocks inhibitory signals, leading to uncontrollable muscle spasms and a particularly unpleasant death. Depressants assist the binding of inhibitory neurotransmitters or (like alcohol) interfere with the action of excitory neurotransmitters. Drugs that relieve pain typically engage with inhibitory receptors. Morphine, the main active ingredient of opium, binds to so-called opioid receptors in the spinal cord, which inhibit the transmission of pain signals to the brain. There are also opioid receptors in the brain itself, which is why morphine and related opiate drugs have a mental as well as a somatic effect. These receptors in the brain are the binding sites of peptide molecules called endorphins, which the brain produces in response to pain. Some of these are themselves extremely powerful painkillers. Cannabinoids, the active ingredients of cannabis, also bind to inhibitory neuroreceptors in the brain to produce pain relief. The natural target of these receptors is a molecule called anandamide, which, like endorphins, is produced in response to pain signals. A closely related molecule called oleamide seems to be the biochemical trigger that induces natural sleep. Not all pain-relieving drugs analgesics work by blocking the pain signal.

Some prevent the signal from ever being sent. Pain signals are initiated by peptides called prostaglandins, which are manufactured and released by distressed cells. Aspirin acetylsalicylic acid latches onto and inhibits one of the enzymes responsible for prostaglandin synthesis, cutting off the cry of pain at its source. Unfortunately, prostaglandins are also responsible for making the mucus that protects the stomach lining, so one of the side effects of aspirin is the risk of ulcer formation. One of the surprising recent discoveries in neuroscience was that extremely small inorganic molecules can also act as neurotransmitters. Carbon monoxide and nitric oxide – both of them two-atom molecules – serve this function. They are both poisonous in large doses, because they compete with oxygen in binding to haemoglobin. But 'the poison is in the dose', and in small amounts nitric oxide does some important things. It triggers the dilation of blood vessels, which can relieve stress on the heart. This is why nitroglycerin, which decomposes to release nitric oxide, is administered to treat heart problems. The improvement in circulation initiated by nitric oxide provides the basis for the drug Viagra, which is used to treat erectile dysfunction in men.

SUPRAMOLECULAR CHEMISTRY

In recent decades, scientists have become interested in mimicking, in synthetic systems, some of the molecular communication processes of the cell. There are many motivations for this. Drug development is often a matter of concocting a good disguise, so that a synthetic molecule will pass itself off as a natural one and bind preferentially to a receptor, blocking or initiating a biochemical signal. Signal transduction in the eye's retinal cells and in the olfactory system suggests the concept of molecular sensors that can detect light or other molecules with high sensitivity. Molecular engineers are looking to the olfactory apparatus for inspiration in designing 'artificial noses' that can identify complex mixtures of molecular components. The principle loudly extolled by nature is the 'lock and key': molecules get together when one fits with the other. To turn such a 'recognition' event into a communication process, the binding event should trigger some change in the receptor that allows it to relay the signal downstream.

In biology this relay process is commonly catalytic: binding turns the receptor into an active enzyme. But the signal might also be passed on in other ways: by the emission of light or release of an electron, for instance, or (as in the case of the acetylcholine receptor) by the creation of an electrochemical potential. Building artificial signal-transduction processes at the molecular scale is a common objective in supramolecular chemistry. From its outset, this discipline was biologically inspired. In the 1960s, the French chemist Jean-Marie Lehn investigated so-called crown ether molecules that would recognize and bind specific metal ions. Lehn was interested in transporting ions such as sodium and potassium across lipid membranes. Although this can be mediated by protein channels and pumps, another strategy is to engulf the ion within a molecule that will 'dissolve' in the fatty interior of the membrane wall. Such molecules exist in nature, and are called ionophores. A typical example is valinomycin, a ring-shaped peptide with a central hole into which a potassium ion will fit. Crown ethers are synthetic mimics of valinomycin: they too are ring-shaped, and will bind a metal in their central cavity. The metal ion is held more or less securely depending on the relative sizes of the ion and the hole. If the hole is too big, the metal 'rattles around' and is only loosely bound; too small, and it will not fit. So crown ethers can be tailored to fit specific metal ions – to display molecular recognition, in other words. By the 1970s, Lehn and others were making synthetic receptor molecules of all shapes and sizes, with cavities designed to accommodate a wide range of inorganic and organic targets.

These 'guest' molecules are held in place by interactions with their hosts that are weak relative to the covalent bonds that hold the atoms together in the molecules themselves. In this way the guests can be picked up and released again. That is how valinomycin works as a metal-ion transporter: it captures the ions on one side of the membrane and lets them go again at the other side. Supramolecular chemistry is essentially about bringing molecules together into loose associations that can be disassembled back into their separate components.

When crown ethers take up a metal ion, they change shape. Alone, they are rather loose, floppy rings, like rubber bands. With a metal ion in their core, they become organized into relatively rigid structures in which the ring contains zigzag-like kinks: a crown, in other words. Shape changes of this sort are common when a receptor binds its target.

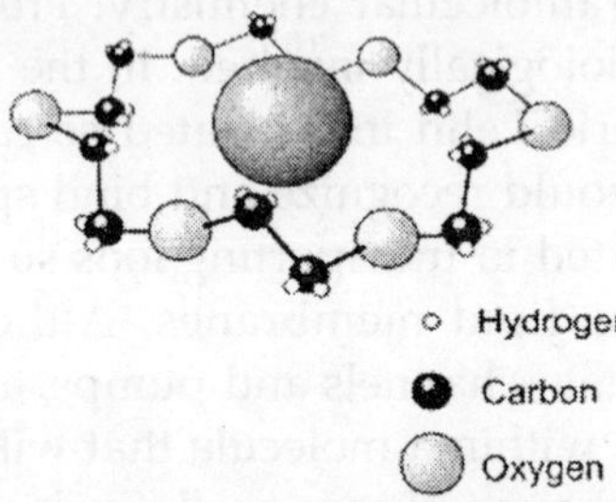

Fig. A crown ether is a cyclic molecule that captures a metal ion in its central cavity

If binding alone is the objective, a big shape change is not terribly desirable, since the internal rearrangements of the receptor make heavy weather of the binding event and may make it harder to achieve. This is why many supramolecular hosts are designed so that they are 'pre-organized' to receive their guests, minimizing the shape change caused by binding.

But if the idea is to use binding as the trigger for relaying some signal, then a shape change is often crucial. A particularly dramatic change in shape induced by host–guest binding was reported in 2000 by Ulrich Koert and colleagues at the Humboldt University in Berlin. They constructed a receptor molecule that can be considered as a series of 'modules': two arms, two legs, and a 'transducer' unit like a flexible torso connecting the arms to the legs. When the two arms close around a zinc ion, the transducer unit flips and pulls the two legs far apart. The legs are tipped with fluorescent groups, which change their emission wavelength – from green to ultraviolet – when the distance between them increases.

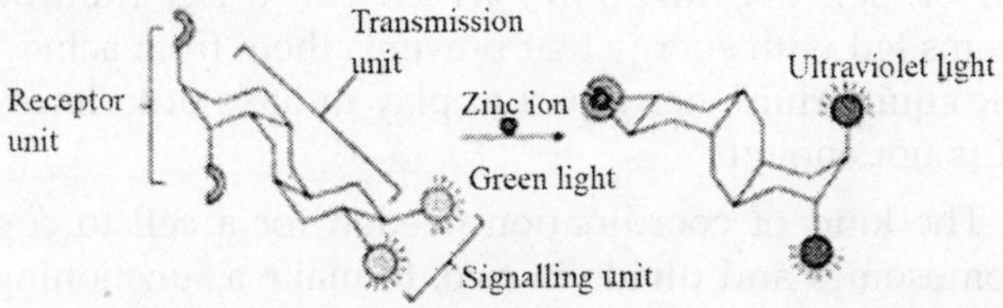

Fig. An artificial transducer molecule that converts binding of a zinc ion into a signalling event, by changing shape and altering its fluorescence properties.

The researchers pointed out how this molecular receptor shows some of the features of a protein receptor involved in signal transduction, responding to binding of the target at one end by altering its shape, and thus its behaviour, at the other end. A shape change that alters a molecule's light-emitting properties has been engineered in several other synthetic receptors. But using recognition and binding to switch a molecule's *catalytic* behaviour, as in the G-protein signalling mechanism, is rather more challenging, since this means ensuring that the final shape is exactly what is needed for the catalyst to do its job. Harder still would be the task of organizing several molecules into a relay that would carry a message downstream. All the same, the skill of supramolecular chemists is increasing daily, and it would not be at all surprising if we were soon to see artificially devised molecular communication systems that approach the sophistication of those the body uses to run its realm harmoniously.

MOLECULAR INFORMATION

It is a necessary but not sufficient characteristic of life that it imposes order on chaos. Chaos is death. If cells cannot send and receive clear messages, if they do things at the wrong time, if their membranes lose their organization, if proteins fail to fold, then life cannot be sustained. We are islands of order in a wild world. Where does this order come from? Organization can arise spontaneously in inorganic matter too: think of the serried ranks of mare's-tail clouds, or the regular ripples in wind-blown sand. It seems highly likely that this kind of 'self-organization', which can arise unbidden in systems fed with energy that prevents them from achieving a static equilibrium, has a part to play in life's orderliness. But that is not enough.

The kind of coordination needed for a cell to copy its chromosomes and divide in two, to make a functioning and reproducible protein molecule, or indeed to grow from a single fertilized egg into a multicelled Mozart, cannot rely on the 'blind' patterning processes that paint the skies and the deserts. There needs to be a firmer hand on the wheel. We have all heard of what this guiding hand consists. It is DNA, a string of molecular beads chopped and bundled into our forty-six little X-shaped chromosomes. The human genome – our full complement of DNA – is often called the 'book of life'.

Scientists have just finished decoding the first draft of this book: they have sketched out in broad detail the molecular messages in each of the chromosomes. Incautious things are said about the project to map the human genome. One hears, for instance, that a sufficiently skilful engineer could make a human from the information therein. This is nonsense. The body is full of molecules that are not encoded in the genome – it encrypts only proteins, and even those in somewhat garbled and incomplete form. The genome tells us nothing about the lipids that make up cell membranes, let alone about how they are driven by physical forces to aggregate into sheets, loops, and spheres. The genes will not tell us how neural signalling works, how the brain encodes thoughts and sensations in delicately timed trains of electrical pulses. There is no gene for bone, for tooth enamel. The genome is the book of the cell in much the same way as the dictionary is the book of a performance of *Waiting for Godot*. It is all in there, but you will not deduce one from the other. Nevertheless, the genome *is* an instruction booklet in molecular form. It tells us how to make proteins, the molecules that orchestrate the spectacular molecular performance of life.

The theoretical biologist John Hopfield of Princeton University points out that this is one of the many ways in which biology provides an 'existence theorem' to inspire chemists. 'Mathematicians use this term', he says, 'in reference to a proof that some function which they want to construct actually does exist, and is not impossible. In this sense, the observation of birds flying provides an existence theorem that an engineer should be able to design a flying machine.' In the same way, genetics shows that it is possible to perform computing with molecules. For computing is nothing but the storage, transmission, and processing of information, all of which the genetic machinery can do. Says Jean-Marie Lehn, 'there is a "molecular logic of living organisms"'.A genuine molecular logic is more precise: it requires not simply that one molecule affects the behaviour of another, but that they can transfer and manipulate encoded information in well-defined ways.

This is how a computer works, by passing data between switches and memory devices made from semiconducting and magnetic materials. Computing with molecules is just one of the ways in which the dimension of information is entering molecular science. More generally, chemists are becoming accustomed to the idea that molecules can be programmed to behave in certain ways: that their properties can be written into the fabric of the molecule just as a set of instructions can be programmed into a robot. Lehn says 'The outlook ... is toward a general science of informed matter.' Such a chemistry is a genuinely new science, in many ways quite different from the traditional chemistry that makes useful substances. It is a science that is about an active 'becoming' rather than a passive 'being'. It is already happening, and we do not know where it might take us.

HOW THE CELL 'BECOMES'

Every book is written in a particular language, and the genome is no different. The language of the genes is a simple code, whose characters are the four nucleotide molecules that represent the beads of DNA's molecular necklace. Each of these molecules contains a so-called base, which encodes the information. There are four DNA bases: adenine, cytosine, guanine, and thymine (A, C, G, and T). As DNA is a linear polymer of nucleotide units, the information it uncodes can be represented as a linear string of these four characters. Part of it might look something like this:

GTGGATTGACATGATAGAAGCACTCTACTATATTC

A four-letter alphabet might seem a rather limited system for writing complex messages. But, if we think of this sequence as a code rather than strictly as an alphabet, it can be as complex as you like. We could, for example, denote every letter in the Roman alphabet by a series of several bases: GTG for 'a', GAT for 'b', and so on. The number of permutations of four characters in groups of three is sixty-four - more than enough to encode all the alphabet. Using such a code, we could write the Bible as a string of A's, G's, C's, and T's. The cell does not have much use for the message of the Bible; what it needs are messages for making proteins. The way that a protein chain folds up is determined by its amino-acid sequence - so the 'information' for making a protein is uniquely specified by this sequence. DNA encodes this information using a cipher just like that suggested: groups of three bases represent each amino acid. This is the genetic code.

How a particular protein sequence determines the way its chain folds is not yet fully understood. This means that we cannot deduce a gene's function from its sequence alone although we can sometimes make good guesses. The first draft of the human genome is full of genes of unknown purpose. Nevertheless, the principle of information flow in the cell is clear. DNA is a manual of information about proteins. We can think of each chromosome as a separate chapter, each gene as a word in that chapter, and each sequential group of three bases in the gene as a character in the word. Proteins are translations of the words into another language, whose characters are amino acids. In general, only when the genetic language is translated can we understand what it means. DNA is a double-stranded polymer: two chains are twisted around one another in a double helix. Each of these strands consists of a string of nucleotides, embodying encoded information. But the strands are not identical. They are stuck together by a kind of zip of hydrogen bonds between the bases on one strand and those on the other.

Although all the bases can form hydrogen bonds, they have distinct preferences in their pairing relationships: A sticks to T, and G to C. So the twin helices of DNA have complementary sequences: wherever an A appears in one strand, a T appears in the other, and so forth. This means that each gene is written in two versions, echoing each other in a kind of mirror language. The particular partnerships of bases are dictated by their shapes. The A and G bases are similar molecules, as are C and T. So an A–T pairing has much the same overall shape and size as a G–C pairing. These pairs point inwards between the two coiled strands like rungs on a spiral staircase.

Because the rungs are the same size, the strands coil evenly. An A paired with a G would create a bulge, and the resulting distortion would destabilize the pairing. Likewise, if C and T were to pair up, there would be a constriction in the double helix. A–C and G–T pairings, meanwhile, are discouraged by the disposition of hydrogen-bonding groups in these base pairs. So the pairing preferences are a matter of a good fit – of *complementarity* – between the two partners. This is one of the crucial points about biological information flow: the transmission of data occurs via molecular-recognition processes, which ensure that each part of the message is read correctly.

DNA is replicated – the genome is copied – when a cell divides. Because the two strands are complementary, each of them can serve as a template on which the other can be put together. If A always binds preferentially to T and so forth, the sequence of a 'naked' single strand will guide unlinked nucleotides to line up in the right order for forming the complementary strand. In order to act as a template, a single strand is unzipped from its partner by special enzymes. Complementary strands are then pieced together along both of the exposed strands; an enzyme called DNA polymerase catalyses the addition of each new nucleotide. So the two new double helices each contain one strand from the original.

Although enzymes help the process along, the essential information needed for the copying process is already written into the DNA templates. In the early 1980s, Leslie Orgel at the Salk Institute in California and co-workers showed that individual nucleotides can be assembled into polymers on a template of complementary nucelotides without assistance from enzymes. A string of eight C-bearing RNA nucleotides, will act as a template for joining together eight G-bearing nucleotides. Orgel had to cheat a little, however, by using G nucleotides that had been 'activated' by the addition of a reactive chemical group, which helped them join together. This template-assisted polymerization is not, in itself, replication: the new strand is complementary to the template, not identical.

The first example of genuine artificial molecular replication was reported in 1986 by German chemist Günter von Kiedrowski. He used the same templating process, but chose a template that was self-complementary: it was its own complement. The template was a six-nucleotide DNA molecule with the sequence CCGCGG. Its complementary sequence is the same as this, because the two strands of the helix line up with one oriented in the reverse direction to the other. Von Kiedrowski assembled this complement from two three-nucleotide fragments, again activated to help them link up.

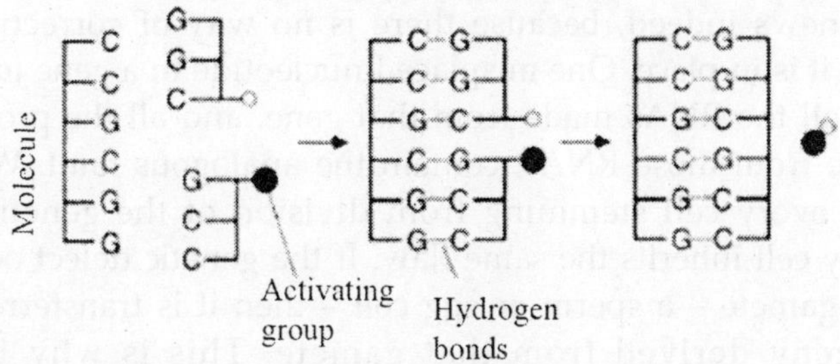

Fig. Molecular replication in a six-unit nucleic acid.

ERRORS AND JUNK

Occasionally a wrong nucleotide or amino acid will be inserted into the chain, for molecules are not capable of perfect recognition all the time. Probably about one in every twenty or so proteins is incorrectly made. Does this matter? On the whole, no. It will be an unusual situation if, between me and the publishers, we spot all the typographical errors in this book before it goes to print. But hopefully none of them will be so serious that you are unable to catch my meaning. Similarly, in proteins most of the chain is scaffolding, which holds in place the few amino-acid residues that conduct the protein's catalytic task. An error here and there in the scaffolding might not be serious. Sometimes the consequence of an error might be a completely non-functional molecule; but because the cell makes not one but typically dozens or hundreds of enzyme molecules for any particular task, one or two duds do not matter. Far more serious are *systematic* errors, which arise further back in the stream of biological information flow - closer to the ultimate repository of information.

A wrongly transcribed RNA molecule could generate hundreds of faulty proteins. So there are enzymes that check the transcription process quite carefully for copying errors, reducing their frequency to around one in ten thousand. Even errors in transcription are seldom a matter of serious consequence: after all, RNA molecules are ephemeral; the cell can always make more of them. But an error in DNA can be bad news indeed, because there is no way of correcting it once it is in place. One misplaced nucleotide in a gene means that all the RNAs made from that gene, and all the proteins made from those RNAs, contain the analogous fault. Worse still, every cell stemming from division of the genetically faulty cell inherits the same flaw. If the genetic defect occurs in a gamete - a sperm or egg cell - then it is transferred to progeny derived from that gamete. This is why DNA replication is scrutinized extremely carefully by 'proof-reading' enzymes, which permit no more than one error in a billion bases to creep through. Without these molecular proof-readers, we would acquire about 1,000 defective genes in every new cell. Inheritable errors, the result of mistakes made in DNA replication during production of the gametes, are known as mutations. Once established, they are passed on from parent to offspring right down the genealogical tree.

Mutations are responsible for genetically related disorders such as cystic fibrosis, as well as for genetically linked predispositions to conditions such as cancer and heart disease. But, in spite of such terrible consequences, mutations are also the spice of life. Indeed, we owe to them our very existence. If the primitive single-celled organisms replicating in the steamy broth of the early Earth had not picked up occasional mutations - if they had invariably copied their DNA without a single mistake - there would have been no evolution, no emergence of complex life. Something else will doubtless have happened to my text when the proofs come back from the publishers. It came as something of a surprise in the mid-1970s to find that genes too need editing. The RNA transcript that peels off from the DNA template is not fit for translation to proteins, for it contains a lot of useless information. These 'primary RNA transcripts' are rather like sentences that have fragments of other sentences inserted seemingly at random. The RNA molecules need heavy editing before they present a clear message fit for translation. The useless inserts are called introns, and sometimes they make up most of a gene. They are also known as non-coding sequences, since they do not encode parts of proteins. Enzymes snip out the introns from the RNA primary transcript, and splice together the two ends of the coding regions called exons. This is just one of the ways in which the supposed 'book of the cell' is mostly garbled nonsense - or boring repetition. It is thought that only about 2-3 per cent of the entire human genome codes for proteins. Some sequences get repeated for good reason. Each human chromosome ends in the sequence TTAGGG repeated about 2,500 times. These sections, called telomeres, are thought to keep the chromosomes stable.

They get shortened each time a cell divides, and their eventual erosion contributes to the ageing process. But many other repeat sequences serve no useful function. Transposons are repeats that jump about in the genome, leaving copies as they go. They are thought to be a genetic parasite living in the very core of our being, whose only purpose is to replicate themselves. Introns may be the remnants of ancient transposons that lost the ability to move on. The protein machinery for cutting, splicing, replicating, and synthesizing nucleic acids provides the principal tools for genetic biotechnology, the manipulation of genomes. Restriction enzymes, for instance, are proteins that can recognize a specific short sequence of DNA and cut the chain at that point. Ligases join loose ends of DNA together. Stretches of DNA can be replicated indefinitely in a test tube using DNA polymerases. The double strands are separated by heating, exposing them for templated replication. Repeated cycles of replication and heating multiply the DNA exponentially. This process, called the polymerase chain reaction (PCR), uses a DNA polymerase taken from a bacterium that lives in hot springs. The enzymes of this bacterium have evolved to withstand high temperatures, and so this DNA polymerase is not destroyed by the heating cycles. These tools allow scientists to 'rewrite the book': to insert new genes into an organism's genome. Crop scientists are interested in developing plants with genes that confer resistance to pests, or to drought, or to particular herbicides, as well as incorporating genes that improve the flavour of the crop, or its growth rate, or whatever.

One potential danger is that genes for herbicide resistance, might become transferred from agricultural crops into weeds, generating new breeds of 'superweed'. The likelihood of this 'horizontal' trans-species transfer of genes is not known. Some people object to genetic engineering on the grounds that it is ethically wrong to tamper with the fundamental material of life - DNA - whether it is in bacteria, humans, tomatoes, or sheep. One can understand such objections, and it would be arrogant to dismiss them as unscientific. Nevertheless, they do sit uneasily with what we now know about the molecular basis of life. The idea that our genetic make-up is sacrosanct looks hard to sustain once we appreciate how contingent, not to say arbitrary, that make-up is. Our genomes are mostly parasite-riddled junk, full of the detritus of over three billion years of evolution. There seems little that is admirable or elegant in this unruly library; rather, the admiration should be reserved for the cohorts of diligent proteins that painstakingly sift snippets of meaning from reams of nonsense. It is truly amazing how well the whole affair works; but, like most of life, it is a makeshift compromise in which efficiency and tidiness count for little.

CONSTRUCTION BLUEPRINTS

DNA is a supreme example of 'informed matter'. It is programmed to assemble in a highly specific manner, each nucleotide marrying up with its complement over thousands of base pairs. This kind of programmed self-assembly represents one of the goals of supramolecular chemistry. Atoms do not in themselves display many powers of discrimination; but by making the molecule, rather than the atom, the fundamental building block, supramolecular chemists are able to programme much more guiding information into their bricks and mortar. Yet DNA provides more than an existence theorem for programmed self-assembly. It can supply the very fabric. Why not use the principles of complementary base pairing to join together DNA girders into structures much more complex than the double helices of the cell? This concept has been explored by Nadrian

Seeman at New York University. He commandeers the enzymatic apparatus of biotechnology to cut and splice DNA into remarkable edifices, such as cagelike polyhedra: a cube and a truncated octahedron. The edges of these structures are DNA double helices, but at the corners three coils meet in a triple junction.

These junctions are cunningly woven: the twin strands go separate ways along different edges, where they intertwine with new strands. Seeman and his coworkers make these junctions by piecing together synthetic DNA with carefully planned sequences. The trick is then to assemble triple junctions into a three dimensional geometric molecular object. Seeman gives the branches 'sticky ends' where one strand extends beyond the other. Here the exposed, unpaired bases are ready to match up with those on another strand - but only if it has a complementary sequence.

Fig. A polyhedral molecular assembly made from double-stranded DNA.

In this way, the ends are selectively sticky and can be assigned one to another, so that the whole structure is programmed to build itself from its component parts. Once the sticky ends are married up to their partners by base-to-base hydrogen bonding, ligation enzymes forge strong bonds to secure the backbone. For the present time, these molecular constructions are follies of virtuosity: demonstrations of the astonishing control that molecular recognition can provide over nanometre-scale architecture. But Seeman suggests that his DNA frameworks might act as scaffolds for assembling other molecules and materials in useful ways. It is possible to coat DNA strands with silver, transforming them into electrically conducting molecular wires. Might we one day build tiny electronic circuits by 'genetically' programming the wires to link up in a specified pattern? Moreover, sticky-ended DNA can clip together molecular-scale objects in a selective manner. Chad Mirkin and colleagues at Northwestern University in Illinois have used this idea to assemble little particles of gold into clusters. The particles are just a few nanometres in size – so-called nanocrystals. Each has a tag consisting of single-stranded DNA, but, because the tag sequences are not complementary, the particles remain separate. By adding lone DNA strands whose two ends complement the sequences of the tags, the researchers are able to link the nanoparticles together. The resulting clusters scatter blue light strongly, so the solution turns the colour of wine.

Mirkin and colleagues are now developing this technique commercially as a method for simple visual detection of DNA strands with a particular sequence – something that is commonly required in genetic analysis. By assembling nanocrystals of metals or semiconductors into organized arrays, some researchers hope to be able to build electronic devices far smaller than those currently made with the conventional microfabrication techniques used for creating silicon chips. Semiconductor nanocrystals could act as memory elements for storing electronic information, and they interact with light in ways that could be useful for making light-based informationprocessing devices. Programming the assembly of nanocrystals using DNA linkers might provide one way of arranging them into circuit patterns. Another possibility has been suggested by the work of Angela Belcher of the University of Texas and co-workers, who used the molecular-recognition properties of proteins rather than DNA. They developed small peptide molecules that would recognize and stick to the surfaces of different kinds of semiconductor. The peptides could 'feel' how the atoms were organized at the surface of the semiconductor crystals. Genetically engineered motor proteins with peptide arms that recognize certain kinds of semiconductor might one day be used to drag nanocrystals around on a molecular building site and arrange them in a circuit pattern.

MOLECULAR LOGIC

Since the invention of the computer in the 1940s, the computing power of new machines has roughly doubled every eighteen months. This trend, known as Moore's law after Gordon Moore, the co-founder of Intel who first pointed it out in 1965, is driven by miniaturization. Computer power increases as it becomes possible to pack more circuit components into a given space. But, if Moore's law is to hold fast for another twenty-five years or so, electronic devices must shrink to nanometre sizes: the scale of molecules. No one yet knows how that will be achieved, for at such scales the silicon transistor – the workhorse of integrated circuits – becomes too leaky a switch. In order to continue making computers faster and more powerful, a growing school of thought says that their components will have to be individual molecules. This is so different a vision from conventional information technology that it would be a bold or reckless speculator who invests in it. Yet it is not a new idea. In 1974 US chemists Mark Ratner and Ari Aviram proposed a design for a single-molecule rectifier a device that passes current in only one direction.

Just a few years later, carbon-based polymers were discovered that can conduct electricity, and researchers began to hope that individual molecules of these materials might supply the wires for a 'molecular computer'. The field of *molecular electronics* was born. But for the next decade or so, nothing much happened. It was an idea before its time, lacking any experimental means to synthesize, arrange, or probe the kinds of molecular devices it dreamed of. In recent years, several paths have converged to create a resurgence in the field, and at last molecular electronics – and its corollary, molecular computing – are starting to gather serious attention from the people who matter: the companies who build computers. One of the central ingredients of a molecular information processing technology is the switch: a device to supplant the transistor. In the most rudimentary terms, a switch can exist in two different stable states – 'on' and 'off'. The transistor passes a current when 'on', and blocks it when 'off'. But a switch is useful for information processing only if it can be hooked up to others, so that the devices can talk to each other and pass information to and fro.

This is hard to achieve with molecules. Yet something of the kind was announced in 1999 by James Heath and co-workers from the University of California at Los Angeles, in collaboration with scientists from the computer giants Hewlett-Packard. They interconnected several switches based on an organic molecule to produce an electrically controlled logic gate. In computer circuits, information is encoded in binary form – as a series of 1's and 0's. A signal of 1 corresponds to a electrical pulse at a certain voltage; a signal of 0 corresponds to zero voltage. Those are the only two signals dispatched through the circuit – there are no signals of $^1/_2$ or 2. Data is encoded in the sequence of 1's and 0's just as DNA encodes information in a sequence of nucleotide bases. The binary code is simpler than the genetic code: it has only two characters. Each unit of information in the coded signal – each 1 and 0 – is called a binary digit, or bit. Computers manipulate binary information and perform calculations using logic gates: devices or circuits that make decisions. A logic gate receives one or more input signals and releases one or more output signals. The outputs depend on what the inputs say. An AND gate, for instance, receives two input bits and produces one output. If both inputs are 1's, the output is also 1; but any other combination of inputs generates an output of 0.

Simple combinations of logic gates like these can perform arithmetic: reading two numbers encoded in binary form and generating an output that encodes their sum (addition) or their difference (subtraction). Heath and colleagues constructed AND gates from a molecular switch called a rotaxane. This is an assembly of two molecules: a ring threaded on a rod. The ring is prevented from falling off by big capping units fixed to the ends of the rod. The rod is designed to attract the ring, so that the two molecules interlock spontaneously when mixed together. The end caps are added afterwards. Heath's colleague Fraser Stoddart at UCLA developed techniques for making molecular assemblies like this while at the University of Sheffield in England in the late 1980s. The researchers arranged rotaxane molecules in a layer on a metal electrode, and deposited fine metal wires on top of them. By applying a voltage to the wires, the molecules could be switched from a low-conductivity to a high-conductivity state. Many thousands of molecules, attached to a single wire, constituted a single, switchable device.

The researchers connected several of these devices together to make an AND gate. In principle, they said, it should be possible to build each device from just a single switchable molecule. But it is difficult to make electrical connections to, and measure tiny currents through, single molecules. Yet even this is not impossible. Mark Reed, James Tour, and co-workers in the USA have measured the electrical conductivity of a single 'molecular wire' connecting two gold electrodes. Fraser Stoddart, in collaboration with Vincenzo Balzani at Bologna and their co-workers, has demonstrated a different logic operation, called XOR, in a single-molecule gate. Like AND, an XOR gate has two inputs and one output. The output signal is 1 if the inputs are different (0,1 or 1,0) and 0 if they are the same (0,0 or 1,1). The researchers observed this behaviour in a pseudorotaxane – a threaded-ring molecular assembly with no end stoppers, so that the ring can slip off the rod. Rather than using electrical signals, the device received chemical inputs and produced an optical output. That is to say, it altered its light-emission behaviour (fluorescence) depending on whether two 'chemical signals' were 'on' or 'off' – whether the two chemicals were present or not.

This is analogous to the way that cellsurface receptor proteins work – they send out some kind of signal depending on whether or not they have bound their target molecules. Using similar principles, A. Prasanna de Silva and Nathan McClenaghan at the University of Belfast in Northern Ireland have combined two molecular logic gates in a way that permits them to conduct elementary arithmetic. They have, in other words, been able to use molecules to count and to perform simple sums, such as 1 + 1 = 2. It is, of course, a long way from adding one plus one to making a computer that compares with silicon-based devices. But studies such as these demonstrate an important principle: molecules can be used for computation, and at the level of one device per molecule. The molecular computer is at last starting to look like more than canny advertising.

Molecular logic conducted by a needle-and-thread molecular assembly called a rotaxane.

The closer we look at the idea, the more we can see similarities with the challenges that the body faces: how to arrange molecules where you want them, how to transmit and amplify signals, how to grow wires between two switching devices (such as neurons), how to cope with errors, how to control the relative timing of events. The computer engineers of the future may need to know a lot of biology.

DNA COMPUTING

As if to drive that message home, in recent years some scientists have shown that computing can be conducted using DNA. This brings us full circle, it it suggested that DNA provides a kind of proof-of-principle for molecular computation. But in the cell it provides the programme for making proteins. No one dreamed, until Leonard Adleman suggested, that DNA could be used to solve the same kinds of problems as computers. Adleman realised that the genetic code can be used, just like the binary code of computer science, to encode mathematical problems. He showed that biotechnological techniques for manipulating and rearranging DNA can be used to generate all possible answers to such a problem, each one encoded in a molecule of DNA. Techniques for analysing DNA sequences are then employed to screen through all these possible answers and identify the correct one. For certain kinds of mathematical problem, computers have no short cuts to the right answer.

They simply have to test out all options, and select the best. If the number of possible answers is large, this search can take a very long time. Such problems are some of the hardest to solve using conventional computers. A classic example is the 'travelling salesman' problem, which entails working out the shortest route connecting a large number of points in space ('cities') so that each is visited just once. Adleman showed that, by shuffling and splicing short segments of DNA at random, all the solutions to these problems may be encoded in a test tube of single-stranded DNA molecules. The number of such solutions might be huge – but the number of molecules in a test tube is greater still. And, because all the possible answers are produced and tested at once, rather than one at a time, in principle DNA computing can find the 'best' answer rapidly. Whether or not DNA computing proves to be useful in a practical sense, it has a strong allegorical appeal. It drives home the message that the molecular basis of life is rooted in the manipulation of information.

It is often said that each age tends to interpret the world through models derived from its most advanced technology, and so maybe in the Age of Information we should be wary of becoming too dogmatic about such a (partial) answer to the perennial question that haunted Haldane, Schrödinger, and countless others. It is perhaps more important that we regard this as a demonstration of the fabulously dynamic, interactive world inhabited, unseen and too often unsung, by molecules.

4

Technology Trends Genomics

By 2015, biotechnology will likely continue to improve and apply its ability to profile, copy, and manipulate the genetic basis of both plants and animal organisms, opening wide opportunities and implications for understanding existing organisms and engineering organisms with new properties. Research is even under way to create new free-living organisms, initially microbes with a minimal genome.

GENETIC PROFILING AND DNA ANALYSIS

DNA analysis machines and chip-based systems will likely accelerate the proliferation of genetic analysis capabilities, improve drug search, and enable biological sensors. The genomes of plants (ranging from important food crops such as rice and corn to production plants such as pulp trees) and animals (ranging from bacteria such as *E. coli*, through insects and mammals) will likely continue to be decoded and profiled. To the extent that genes dictate function and behaviour, such extensive genetic profiling could provide an ability to better diagnose human health problems, design drugs tailored for individual problems and system reactions, better predict disease predispositions, and track disease movement and development across global populations, ethnic groups, and other genetic pools. Note that a link between genes and function is generally accepted, but other factors such as the environment and phenotype play important modifying roles. Gene therapies will likely continue to be developed, although they may not mature by 2015. Genetic profiling could also have a significant effect on security, policing, and law. DNA identification may complement existing biometric technologies (e.g., retina and fingerprint identification) for granting access to secure systems (e.g., computers, secured areas, or weapons), identifying criminals through DNA left at crime scenes, and authenticating items such as fine art.

Genetic identification will likely become more commonplace tools in kidnapping, paternity, and fraud cases. Biosensors some genetically engineered may also aid in detecting biological warfare threats, improving food and water quality testing, continuous health monitoring, and medical laboratory analyses. Such capabilities could fundamentally change the way health services are rendered by greatly improving disease diagnosis, understanding predispositions and improving monitoring capabilities. Such profiling may be limited by technical difficulties in decoding some genomic segments and in understanding the implications of the genetic code. Our current technology can decode nearly all of the entire human gene sequence, but errors are still an issue, since Herculean efforts are required to decode the small amount of remaining sequences. More important, although there is a strong connection between an organism's function and its genotype, we still have large gaps in understanding the intermediate steps in copying, transduction, isomer modulation, activation, immediate function, and this function's effect on larger systems in the organism.

Proteomics the study of protein function and genes is the next big technological push after genomic decoding. Progress may likely rely on advances in bioinformatics, genetic code combination and sequencing akin to hierarchical programming in computer languages, and other related information technologies. Despite current optimism, a number of technical issues and hurdles could moderate genomics progress by 2015. Incomplete understanding of sequence coding, transduction, isomer modulation, activation, and resulting functions could form technological barriers to wide engineering successes. Extensive rights to own genetic codes may slow research and ultimately the benefits of the decoding. At the other extreme, the inability to secure patents from sequencing efforts may reduce commercial funding and thus slow research and resulting benefits. In addition, investments in biotechnology have been cyclic in the past. As a result, advancements in research and development (R and D) may come in surges, especially in areas where the time to market and thus time to return on investment is long.

CLONING

Artificially producing genetically identical organisms through cloning will likely be significant for engineered crops, livestock, and research animals. Cloning may become the dominant mechanism for rapidly bringing engineered traits to market, for continued maintenance of these traits, and for producing identical organisms for research and production. Research will likely continue on human cloning in unregulated parts of the world with possible success by 2015, but ethical and health concerns will likely limit wide-scale cloning of humans in regulated parts of the world. Individuals or even some states may also engage in human or animal cloning, but it is unclear what they may gain through such efforts. Cloning, especially human cloning, has already generated significant controversies across the globe. Concerns include moral issues, the potential for errors and medical deficiencies of clones, questions of the ownership of good genes and genomes, and eugenics. Although some attempts at human cloning are possible by 2015, legal restrictions and public opinion may limit their extent.

Fringe groups, however, may attempt human cloning in advance of legislative restrictions or may attempt cloning in unregulated countries. See, the human cloning programme announced by Clonaid. Although expert opinions vary regarding the current feasibility of human cloning, at least some technical hurdles for human cloning will likely need to be addressed for safe, wide-scale use. "Attempts to clone mammals from single somatic cells are plagued by high frequencies of developmental abnormalities and lethality".

Even cloned plant populations exhibit "substantial developmental and morphological irregularities". Research will need to address these abnormalities or at the very least mitigate their repercussions. Some believe, however, that human cloning may be accomplished soon if the research organization accepts the high lethality rate for the embryo and the potential generation of developmental abnormalities.

GENETICALLY MODIFIED ORGANISMS

Beyond profiling genetic codes and cloning exact copies of organisms and microorganisms, biotechnologists can also manipulate the genetic code of plants and animals and will likely continue efforts to engineer certain properties into life forms for various reasons. Traditional techniques for genetic manipulation (such as cross-pollination, selective breeding, and irradiation) will likely continue to be extended by direct insertion, deletion, and modification of genes through laboratory techniques. Targets include food crops, production plants, insects, and animals. Desirable properties could be genetically imparted to genetically engineered foods, potentially producing: improved taste; ultra-lean meats with reduced "bad" fats, salts, and chemicals; disease resistance; and artificially introduced nutrients (socalled "nutraceuticals"). Genetically modified organisms (GMOs) can potentially be engineered to improve their physical robustness, extend field and shelf life, tolerate herbicides, grow faster, or grow in previously unproductive environments (e.g., in high-salinity soils, with less water, or in colder climates). Beyond systemic disease resistance, *in vivo* pesticide production has already been demonstrated (e.g., in corn) and could have a significant effect on pesticide production, application, regulation, and control with targeted release. Likewise, organisms could be engineered to produce or deliver drugs for human disease control. Cow mammary glands might be engineered to produce pharmaceuticals and therapeutic organic compounds; other organisms could be engineered to produce or deliver therapeutics (e.g., the so-called "prescription banana").

If accepted by the population, such improved production and delivery mechanisms could extend the global production and availability of these therapeutics while providing easy oral delivery. In addition to food production, plants may be engineered to improve growth, change their constitution, or artificially produce new products. Trees will likely be engineered to optimize their growth and tailor their structure for particular applications such as lumber, wood pulp for paper, fruiting, or carbon sequestering (to reduce global warming) while reducing waste byproducts. Plants might be engineered to produce bio-polymers (plastics) for engineering applications with lower pollution and without using oil reserves. Bio-fuel plants could be tailored to minimize polluting components while producing additives needed by the consuming equipment. Genetic engineering of microorganisms has long been accepted and used. Engineering of bacterial properties into plants and animals for disease resistance will likely occur. Other animal manipulations could include modification of insects to impart desired behaviours, provide tagging (including GMO tagging), or prevent physical uptake properties to control pests in specific environments to improve agriculture and disease control. Research on modifying human genes has already begun and will likely continue in a search for solutions to genetically based diseases. Although slowed by recent difficulties, gene therapy research will likely continue its search for useful mechanisms to address genetic deficiencies or for modulating physical processes such as beneficial protein production or control mechanisms for cancer.

Advances in genetic profiling may improve our understanding and selection of therapy techniques and provide breakthroughs with significant health benefits. Some cloning of humans will be possible by 2015, but legal restrictions and public opinion may limit its actual extent. Controls are also likely for human modifications (e.g., clone-based eugenic modifications) for nondisease purposes. It is possible, however, that technology will enable genetic modifications for hereditary conditions (i.e., sickle cell anemia) through *in vitro* techniques or other mechanisms. GMOs are also having a large effect on the scientific community as an enabling technology. Not only do "knock-out" animals (animals with selected DNA sequences removed from their genome) give scientists another tool to study the effect of the removed sequence on the animal, they also enable subsequent analysis of the interaction of those functions or components with the animal's entire system. Although knock-outs are not always complete, they provide another important tool to confirm or refute hypotheses regarding complex organisms.

BROADER ISSUES AND IMPLICATIONS

Extant capabilities in genomics have already created opportunities yet have generated a number of issues. As more organisms are decoded and the functional impli- cations of genes are discovered, concerns about property and privacy rights for the sequencing will likely continue. The ability to profile an individual's DNA is already raising concerns about privacy and excessive monitoring. Examples include databases of DNA signatures for use in criminal investigations, and the potential use of genetically based health predispositions by insurance companies or employers to deny coverage or to discriminate. The latter may raise policy issues regarding acceptable and unacceptable profiling for insurance or employment. This issue is further worrisome because the exact code-tofunction mechanisms that trigger many disease predispositions are not well understood. Issues may also arise if a strong genetic basis of human physical or cognitive ability is discovered. On the positive side, understanding a person's predisposition for certain abilities (or limitations) could enable custom educational or remediation programmes that will help to compensate for genetic inclinations, especially in early years when their effect can be optimized.

On the negative side, groups may use such analyses in arguments to discriminate against target populations, aggravating social and international conflicts. Although the genetic profiles of plants have been modified for centuries using traditional techniques, questions regarding the safety of genetically modified foods have sparked international concerns in the United Kingdom and Europe, forcing a campaign by biotechnology companies to argue the safety of the technology and its applications. Some have argued that genetic engineering is actually as safe or safer than traditional combinatorial techniques such as irradiated seeds, since there often is strong supporting information concerning the function of the inserted sequences. Governments have been forced into the issue, resulting in education efforts, food labeling proposals, and heated international trade discussions between the United States and Europe on the importation of GMOs and their seedlings. As genetic modification becomes more common, it may become more difficult to label and separate GMOs, resulting in a forcing function to resolve the issue of how far the technology should be applied and whether separate markets can be maintained in a global economy. This debate is starting to have global effects as populations in other countries begin to notice the impassioned debates in the United Kingdom and Europe. Some have likened the anti-biotechnology movement to the anti-nuclear-power movement in scope and tactics, although the low cost and wide availability of basic genomic equipment and know-how will likely allow practically any country, small business, or even individual to participate in genetic engineering (Hapgood, 2000 [40]). Such wide technology availability and low entry costs could make it impossible for any movement or government to control the spread and use of genomic technology.

At an extreme, successful protest pressures on big biotechnology companies together with wide technology availability could ultimately drive genomic engineering "underground" to groups outside such pressures and outside regulatory controls that help ensure safe and ethical uses. This could ironically facilitate the very problems that the anti-biotechnology movement is hoping to prevent. Cloning and genetic modification also raise biodiversity concerns. Standardization of crops and livestock have already increased food supply vulnerabilities to diseases that can wipe out larger areas of production. Genetic modification may increase our ability to engineer responses to these threats, but the losses may still be felt in the production year unless broad-spectrum defences are developed. In addition to food safety, the ability to modify biological organisms holds the possibility of engineered biological weapons that circumvent current or planned countermeasures. On the other hand, genomics could aid in biological warfare defence (e.g., through improved understanding and control of biological function both in and between pathogens and target hosts as well as improved capability for engineered biosensors). Advances in genomics, therefore, could advance a race between threat engineering and countermeasures. Thus, although genetic manipulation is likely to result in medical advances, it is unclear whether we will be in a safer position in the future. The rate at which GMO benefits are felt in poorer countries may depend on the costs of using patented organisms, marketing demands and approaches, and the rate at which crops become ubiquitous and inseparable from unmodified strains. Consider, current issues related to human immunodeficiency virus (HIV) drug development and dissemination in poorer countries. Patentability has fueled research investments, but many poorer countries with dire needs cannot afford the latest drugs and must wait for handouts or patent expiration.

Globalization, however, may fuel dissemination as multinational companies invest in food production across the globe. Also, the rewards from opening previously unproductive land for production may provide the financial incentive to pay the premium for GMOs. Furthermore, widely available genomic technology could allow academics, nonprofit small businesses, and developing countries to develop GMOs to alleviate problems in poorer regions; larger biotechnology companies will focus on markets requiring capital-intensive R and D. Finally, moral issues may play a large role in modulating the global effect of genomics trends. Some people simply believe it is improper to engineer or modify biological organisms using the new techniques. Unplanned side effects (e.g., the imposition of arthritis in current genetically modified pigs) will likely support such opposition. Others are concerned with the real danger of eugenics programmes or of the engineering of dangerous biological organisms.

THERAPIES AND DRUG DEVELOPMENT TECHNOLOGY

Beyond genetics, biotechnology will likely continue to improve therapies for preventing and treating disease and infection. New approaches might block a pathogen's ability to enter or travel in the body, leverage pathogen vulnerabilities, develop new countermeasure delivery mechanisms, or modulate or augment the immune response to recognizing new pathogens. These therapies may counter the current trend of increasing resistance to extant antibiotics, reshaping the war on infections. In addition to addressing traditional viral and bacterial problems, therapies are being developed for chemical imbalances and modulation of chemical stasis. Antibodies are being developed that attack cocaine in the body and may be used to control addiction.

Such approaches could have a significant effect on modifying the economics of the global illegal drug trade while improving conditions for users. Drug development will likely be aided by various technology trends and enablers. Computer simulations combined with proliferating trends for molecular imaging technologies (e.g., atomic-force microscopes, mass spectroscopy, and scanning probe microscopes) may continue to improve our ability to design molecules with desired functional properties that target specific receptors, binding sites, or markers, complementing combinatorial drug search with rational drug design. Simulations of drug interactions with target biological systems could become increasing useful in understanding drug efficacy and safety. Dennis Noble's complex virtual heart simulation has already contributed to U.S. Food and Drug Administration (FDA) approval of a cardiac drug by helping to understand the mechanisms and significance of an effect noticed in the clinical trial. For some better understood systems such as the heart, this approach may become a dominant complement to clinical drug trials by 2015, whereas other more complex systems (e.g., the brain) will likely require more research on the system function and biology.

BROADER ISSUES AND IMPLICATIONS

R and D costs for drug development are currently extremely high and may even be unsustainable, with averages of approximately $600 million per drug brought to market. These costs may drive the pharmaceutical industry to invest heavily in technology advances with the goal of long-term viability of the industry. Combined with genetic profiling, drug development tailored to genotypes, chemical simulation and engineering programmes, and drug testing simulations may begin to change pharmaceutical development from a broad application trial-and-error approach to custom drug development, testing, and prescription based on a deeper understanding of subpopulation response to drugs. This understanding may also rescue drugs previously rejected because of adverse reactions in small populations of clinical trials. Along with the potential for improving success rates, reducing trial costs, and opening new markets for narrowly targeted drugs, tailoring drugs to subpopulations will also have the opposite effect of reducing the size of the market for each drug. Thus, the economics of the pharmaceutical and health industries will likely change significantly if these trends come to fruition. Note that patent protection is not uniformly enforced across the globe for the pharmaceutical industry. As a result, certain regions (e.g., Asia) may continue to focus on production of non-legacy (generic) drugs, and other regions (e.g., the United States, United Kingdom, and Europe) will likely continue to pursue new drugs in addition to such low-margin pharmaceuticals.

BIOMEDICAL ENGINEERING

Multidisciplinary teaming is accelerating advances and products in biomedical engineering and technology of organic and artificial tissues, organs, and materials.

Organic Tissues and Organs

Advances in tissue and organ engineering and repair are likely to result in organic and artificial replacement parts for humans. New advances in tissue regeneration and repair continue to improve our ability to resolve health problems within our bodies. The field of tissue engineering, which is barely a decade old, has already led to engineered commercial skin products for wound treatment. Growth of cartilage for repair and replacement is at the stage of clinical testing, and treatment of heart disease via growth of functional tissue by 2015 is a realistic goal. These advances will depend upon improved biocompatible (or bioabsorbable) scaffold materials, development of 3D vascularized tissues and multicellular tissues, and an improved understanding of the *in vivo* growth process of cellular material on such scaffolds. Research and applications of stem cell therapies will likely continue and expand, using these unspecialized human cells to augment or replace brain or body functions, organs (e.g., heart, kidney, liver, pancreas), and structures. As the most unspecialized stem cells are found in early stage embryos or fetal tissue, an ethical debate is ensuing regarding the use of stem cells for research and therapy.

Alternatives such as the use of adult human stem cells or stem cell culturing may ultimately produce large-scale cell supplies with reduced ethical concerns. Current debates have limited U.S. government funding for stem cell research, but the potential has attracted substantial private funding. Xenotransplantations (transplantation of body parts from one species to a different species) could be improved, aided by attempts to genetically modify donor tissue and organ antibodies, complements, and regulatory proteins to reduce or eliminate rejection. Baboons or pigs, may be genetically modified and cloned to produce organs for human transplant, although large-scale success may not occur by 2015. Beyond rejection, the significance of xenotransplants is likely to be modulated by concerns that diseases such as retro viruses might jump from animals to people as a result of the transplantation techniques. Ethical (e.g., animal rights) and moral concerns as well as possible patenting issues may also result in regulations and limitations on xenotransplants, limiting their significance.

ARTIFICIAL MATERIALS, ORGANS, AND BIONICS

In addition to organic structures, advances are likely to continue in engineering artificial tissues and organs for humans. Multi-functional materials are being developed that provide both structure and function or that have different properties on different sides, enabling new applications and capabilities. Polymers with a hydrophilic shell around a hydrophobic core (biomimetic of micelles) can be used for timed release of hydrophobic drug molecules, as carriers for gene therapy or immobilized enzymes, or as artificial tissues. Sterically stabilized polymers could also be used for drug delivery. Other materials are being developed for various biomedical applications. Fluorinated colloids, are being developed that take advantage of the high electronegativity of fluorine to enhance *in vivo* oxygen transport (as a blood substitute during surgery) and for drug delivery. Hydrogels with controlled swelling behaviour are being developed for drug delivery or as templates to attach growth materials for tissue engineering. Ceramics such as bioactive calcia-phosphate-silica glasses (gel-glasses), hydroxyapetite, and calcium phosphates can serve as templates for bone growth and regeneration. Bioactive polymers (e.g., polypeptides) can be applied as meshes, sponges, foams, or hydrogels to stimulate tissue growth. Coatings and surface treatments are being developed to increase biocompatibility of implanted materials.

Blood substitutes may change the blood storage and retrieval systems while improving safety from blood-borne infections. New manufacturing techniques and information technology are also enabling the production of biomedical structures with custom sizing and shape. It may become commonplace to manufacture custom ceramic replacement bones for injured hands, feet, and skull parts by combining computer tomography and "rapid prototyping" to reverse engineer new bones layer by layer. Beyond structures and organs, neural and sensor prosthetics could begin to become significant by 2015. Retinas and cochlear implants, bypasses of spinal and other nerve damage, and other artificial communications and stimulations may improve and become more commonplace and affordable, eliminating many occurrences of blindness and deafness. This could eliminate or reduce the effect of serious handicaps and change society's response from accommodation to remediation.

BIOMIMETICS AND APPLIED BIOLOGY

Recent techniques such as functional brain imaging and knock-out animals are revolutionizing our endeavors to understand human and animal intelligence and capabilities. These efforts should, by 2015, make significant inroads in improving our understanding of phenomena such as false memories, attention, recognition, and information processing, with implications for better understanding people and designing and interfacing artificial systems such as autonomous robots and information systems. Neuromorphic engineering (which bases its architecture and design principles on those of biological nervous systems) [7] has already produced novel control algorithms, vision chips, head-eye systems, and biomimetic autonomous robots. Although not likely to produce systems with wide intelligence or capabilities similar to those of higher organisms, this trend may produce systems by 2015 that can robustly perform useful functions such as vacuuming a house, detecting mines, or conducting autonomous search.

SURGICAL AND DIAGNOSTIC BIOTECHNOLOGY

Biotechnology and materials advances are likely to continue producing revolutionary surgical procedures and systems that will significantly reduce hospital stays and cost and increase effectiveness. New surgical tools and techniques and new materials and designs for vesicle and tissue support will likely continue to reduce surgical invasiveness and offer new solutions to medical problems. Techniques such as angioplasty may continue to eliminate whole classes of surgeries; others such as laser perforations of heart tissue could promote regeneration and healing. Advances in laser surgery could refine techniques and improve human capability, especially as costs are reduced and experience spreads. Hybrid imaging techniques will likely improve diagnosis, guide human and robotic surgery, and aid in basic understanding of body and brain function. Finally, collaborative information technology (e.g., "telemedicine") will likely extend specialized medical care to remote areas and aid in the global dissemination of medical quality and new advances.

BROADER ISSUES AND IMPLICATIONS

By 2015, one can envision: effective localized, targeted, and controlled drug delivery systems; long-lived implants and prosthetics; and artificial skin, bone, and perhaps heart muscle or even nerve tissue. A host of social, political, and ethical issues such as those discussed will likely accompany these developments. Biomedical advances (combined with other health improvements) are already increasing human life span in countries where they are applied. New advances by 2015 are likely to continue this trend, accentuating issues such as shifts in population age demographics, financial support for retired persons, and increased health care costs for individuals. Advances, however, may improve not only life expectancy but productivity and utility of these individuals, offsetting or even overcoming the resulting issues. Many costly and specialized medical techniques are likely to initially benefit citizens who can afford better medical care; wider global effects may occur later as a result of traditional trickle-down effects in medicine. Some technologies (e.g., telemedicine) may have the opposite trend where low-cost technologies may enable cost-effective consulting with specialists regardless of location. However, access to technology may greatly mediate this dispersal mechanism and may place additional demands on technology upgrades and education.

Countries that remain behind in terms of technological infrastructures may miss many of these benefits. Theological debates have also raised concerns about the definition of what constitutes a human being, since animals are being modified to produce human organs for later xenotransplantation in humans. Genetic profiling may help to inform this debate as we understand the genetic differences between humans and animals. Improved understanding of human intelligence and cognitive function could have broader legal and social effects. An understanding of false memories and how they are created could have an effect on legal liabilities and courtroom testimony. Understanding innate personal capabilities and job performance requirements could help us determine who would make better fighter pilots, who has an edge in analyzing complex images, and what types of improved training could improve people's capabilities to meet the special demands of their chosen careers. Ethical concerns could arise concerning discrimination against people who lack certain innate skills, requiring objective and careful measures for hiring and promotion.

Eventually, neural and sensory implants (combined with trends toward pervasive sensors in the environment and increased information availability) could radically change the way people sense, perceive, and interact with natural and artificial envi- ronments. Ultimately, these new capabilities could create new jobs and functions for people in these environments. Such innovations may first develop for individuals with particularly challenging and critical functions (e.g., soldiers, pilots, and controllers), but innovations may first develop in other quarters (e.g., for entertainment or business functions), given recent trends. Initial research indicates the feasibility of such implants and interactions, but it is unclear whether R and D and investments will accelerate enough to realise even such early applications by 2015. Current trends have concentrated on medical prosthetics where research prototypes are already appearing so it appears likely that globally significant systems will appear in this domain first.

THE PROCESS OF MATERIALS ENGINEERING

New materials can often be critical enabling drivers for new systems and applications with significant effects. However, it may not be obvious how enabling materials affect more observable trends and applications. A common process model from materials engineering can help to show how materials appear likely to break previous barriers in the process that ultimately results in applications with potential global benefits.

Developments in materials science and engineering result from interdisciplinary materials research. This development can be conveniently represented by the schematic description of the materials engineering process from concept to product/ application. This process view is a common approach in materials research circles and similar representations may be found in the literature. Current trends in materials research that could result in global effects by 2015 are categorized below according to the process description of provides an example of the development process in the area of electro active polymers for robotic devices and artificial muscles.

Concept/Materials Design

Biomimetics is the design of systems, materials, and their functionality to mimic nature. Current examples include layering of materials to achieve the hardness of an abalone shell or trying to understand why spider silk is stronger than steel. *Combinatorial materials design* uses computing power (sometimes together with massive parallel experimentation) to screen many different materials possibilities to optimize properties for specific applications (e.g., catalysts, drugs, optical materials).

Materials Selection, Preparation, and Fabrication

Composites are combinations of metals, ceramics, polymers, and biological materials that allow multi-functional behaviour. One common practice is reinforcing polymers or ceramics with ceramic fibres to increase strength while retaining light weight and

Materials Engineering Process Applied to Electroactive Polymers avoiding the brittleness of the monolithic ceramic. Materials used in the body often combine biological and structural functions (e.g., the encapsulation of drugs). *Nanoscale materials*, i.e., materials with properties that can be controlled at submicrometer ($<10^{-6}$ m) or nanometer (10^{-9} m) level, are an increasingly active area of research because properties in these size regimes are often fundamentally different from those of ordinary materials. Examples include carbon nanotubes, quantum dots, and biological molecules. These materials can be prepared either by purification methods or by tailored fabrication methods.

PROCESSING, PROPERTIES, AND PERFORMANCE

These areas are inextricably linked to each other: Processing determines properties that in turn determine performance. Moreover, the sensitivity of instrumentation and measurement capability is often the enabling factor in optimizing processing, as for nanotechnology and microelectromechanical systems (MEMS). *Rapid prototyping* is the capability to combine computer-assisted design and manufacturing with rapid fabrication methods that allow inexpensive part production (as compared to the cost of a conventional production line). Rapid prototyping enables a company to test several different inexpensive prototypes before committing infrastructure investments to an approach. Combined with manufacturing system improvements to allow flexibility of approach and machinery, rapid prototyping can lead to an *agile manufacturing* capability. Alternatively, the company can use its virtual capability to design and then outsource product manufacturing, thus offloading capital investment and risk.

This capability is synergistic with the information technology revolution in the sense that it is a further factor in globalizing manufacturing capability and enabling organizations with less capital to have a significant technological effect. For the Department of Defence (DoD), it could reduce or eliminate requirements for warehousing large amounts of spares and could enable the Air Force to "fly before they buy." *Self-assembly* refers to the use in materials processing or fabrication of the tendency of some materials to organize themselves into ordered arrays (e.g., colloidal suspensions). This provides a means to achieve structured materials "from the bottom up" as opposed to using manufacturing or fabrication methods such as lithography, which is limited by the measurement and instrumentation capabilities of the day. Organic polymers have been tagged with dye molecules to form arrays with lattice spacing in the visible optical wavelength range and that can be changed through chemical means.

This provides a material that fluoresces and changes colour to indicate the presence of chemical species. *Manufacturing with DNA* might represent the ultimate biomimetic manufacturing scheme. It consists of "functionalizing small inorganic building blocks with DNA and then using the molecular recognition processes associated with DNA to guide the assembly of those particles or building blocks into extended structures". Using this approach, Mirkin and colleagues demonstrated a highly selective and sensitive DNA-based chemical assay method using 13 nm diameter gold parti-cles with attached DNA sequences.

This approach is compatible with the commonly used polymerase chain reaction (PCR) method of amplification of the amount of the target substance.*Micro- and nano-fabrication* methods include lithography of coupled micro- or nano-scale devices on the same semiconductor or biological material. It is important to note the crucial role played in the development of these techniques by the parallel development of instrumentation and measurement devices such as the Atomic Force Microscope (AFM) and the various Scanning Probe Microscopes (SPMs).

PRODUCT/APPLICATION

The trends described will likely work in concert to provide materials engineers with the capability to design and produce advanced materials that will be:

- *Smart*—Reactive materials combining sensors and actuators, perhaps together with computers, to enable response to environmental conditions and changes thereof. (Note, however, that limitations include the sensitivity of sensors, the performance of actuators, and the availability of power sources with required magnitude compatible with the desired size of the system.) An example might be robots that mimic insects or birds for applications such as space exploration, hazardous materials location and treatment, and unmanned aerial vehicles (UAVs).
- *Multi-functional*—MEMS and the "lab-on-a-chip" are excellent examples of systems that combine several functions. Another example is a drug delivery system using a hydrogel with hydrophilic exterior and hydrophobic interior. Consider also aircraft skins fabricated from radar-absorbing materials that incorporate avionic links and the ability to modify shape in response to airflow.
- *Environmentally compatible or survivable*—The development of composite materials and the ability to tailor materials at the atomic level will likely provide opportunities to make materials more compatible with the environments in which they will be used. Examples might include prosthetic devices that serve as templates for the growth of natural tissue and structural materials that strengthen during service (e.g., through temperature- or stress-induced phase changes).

SMART MATERIALS TECHNOLOGY

Several different types of materials exhibit sensing and actuation capabilities, including ferroelectrics (exhibiting strain in response to a electric field), shape-memory alloys (exhibiting phase transition-driven shape change in response to temperature change), and magnetostrictive materials (exhibiting strain in response to a magnetic field). These effects also work in reverse, so that these materials, separately or together, can be used to combine sensing and actuation in response to environmental conditions. They are currently in widespread use in applications from ink-jet printers to magnetic disk drives to anti-coagulant devices. An important class of smart materials is composites based upon lead zirconate titanate (PZT) and related ferroelectric materials that allow increased sensitivity, multiple frequency response, and variable frequency. An example is the "Moonie" – a PZT transducer placed inside a half-moon-shaped cavity, which provides substantial amplification of the response. Another example is the use of composites of barium strontium titanate and non-ferroelectric materials that provide frequency-agile and field-agile responses. Applications include sensors and actuators that can change their frequency either to match a signal or to encode a signal. Ferroelectrics are already in use as nonvolatile memory elements for smart cards and as active elements in smart skis that change shape in response to stress. Another important class of materials is smart polymers (e.g., ionic gels that deform in response to electric fields). Such electro-active polymers have already been used to make "artificial muscles".

Currently available materials have limited mechanical power, but this is an active research area with potential applications to robots for space exploration, hazardous duty of various types, and surveillance. Hydrogels that swell and shrink in response to changes in pH or temperature are another possibility; these hydrogels could be used to deliver encapsulated drugs in response to changes in body chemistry (e.g., insulin delivery based upon glucose concentration). Another variation on this trend for controlled release of drugs is materials with hydrophilic exterior and hydrophobic interior.

BROADER ISSUES AND IMPLICATIONS

A world with pervasive, networked sensors and actuators (e.g., on and part of walls, clothing, appliances, vehicles, and the environment) promises to improve, optimize, and customize the capability of systems and devices through availability of information and more direct actuation. Continuously available communication capability, ability to catalog and locate tagged personal items, and coordination of support functions have been espoused as benefits that may begin to be realised by 2015. The continued development of small, low-profile biometric sensors, coupled with research on voice, handwriting, and fingerprint recognition, could provide effective personal security systems.

These could be used for identification by police/military and also in business, personal, and leisure applications. Combined with today's information technologies, such uses could help resolve nagging security and privacy concerns while enabling other applications such as improved handgun safety (through owner identification locks) and vehicle theft control. Other potential applications of smart materials that would be enabled by 2015 include: clothes that respond to weather, interface with information systems, monitor vital signs, deliver medicines, and automatically protect wounds; airfoils that respond to airflow; buildings that adjust to the weather; bridges and roads that sense and repair cracks; kitchens that cook with wireless instructions; virtual reality telephones and entertainment centres; and personal medical diagnostics (perhaps inter- faced directly with medical care centres).

The level of development and integration of these technologies into everyday life will probably depend more on consumer attitudes than on technical developments. In addition to the surveillance and identification functions mentioned under smart materials above, developments in robotics may provide new and more sensitive capabilities for detecting and destroying explosives and contraband materials and for operating in hazardous environments.

Increases in materials performance, both for power sources and for sensing and actuation, as well as integration of these functions with computing power, could enable these applications. Such trend potentials are not without issues. Pervasive sensory information and access to collected data raise significant privacy concerns. Also, the pace of development will likely depend on investment levels and market drivers. In many cases the immediate benefits and cost savings from smart material applications will continue to drive development, but more exotic materials research may depend on public commitment to research and belief in investing in longer-term rewards.

SELF-ASSEMBLY TECHNOLOGY

Examples of self-assembling materials include colloidal crystal arrays with mesoscale (50–500 nm) lattice constants that form optical diffraction gratings, and thus change colour as the array swells in response to heat or chemical changes. In the case of a hydrogel with an attached side group that has molecular recognition capability, this is a chemical sensor. Self-assembling colloidal suspensions have been used to form a light-emitting diode (nanoscale), a porous metal array (by deposition followed by removal of the colloidal substrate), and a molecular computer switch. The DNA-based self-assembly mentioned was achieved by attaching non-linking DNA strands to metal nanoparticles and adding a linking agent to form a DNA lattice. This can be turned into a biosensor or a nanolithography technique for biomolecules.

BROADER ISSUES AND IMPLICATIONS

Development of self-assembly methods could ultimately provide a challenge to topdown lithography approaches and molecular manufacturing approaches. As a result, it could define the next manufacturing methodology at some time beyond 2015.

RAPID PROTOTYPING TECHNOLOGY

This manufacturing approach integrates computer-aided design (CAD) with rapid forming techniques to rapidly create a prototype (sometimes with embedded sen- sors) that can be used to visualize or test the part before making the investment in tooling required for a production run. Originally, the prototypes were made of plastic or ceramic materials and were not functional models, but now the capability exists to make a functional part, e.g., out of titanium. The discussion of reverse-engineered bones in the section on biomedical engineering.

BROADER ISSUES AND IMPLICATIONS

As discussed, agile manufacturing systems are envisioned that can connect the customer to the product throughout its life cycle and enable global business enterprises. An order would be processed using a computer-aided design, the manufacturing system would be configured in real time for the specific product (e.g., model, style, colour, and options), raw materials and components would be acquired just in time, and the product would be delivered and tracked throughout its life cycle (including maintenance and recycling with identification of the customer). Components of the business enterprise could be dynamically based in the most cost-effective locations with all networked together globally. The growth of this type of business enterprise could accelerate business globalization.

BUILDINGS

Research on composite materials, waste management, and recycling has reached the stage where it is now feasible to construct buildings using materials fabricated from significant amounts of indigenous waste or recycled material content. These approaches are finding an increasing number of cost-effective applications, especially in developing countries. Examples include the Petronas Twin Towers in Kuala Lumpur, Malaysia. These towers are the tallest buildings on earth and are made with reinforced concrete rather than steel. A roofing material used in India is made of natural fibre and agro-industrial waste. Prefabricated composite materials for home construction have also been developed in the United States, and a firm in the Netherlands is developing a potentially ubiquitous, inexpensive housing approach targeted for developing countries that uses spray-forming over an inflatable air shell.

TRANSPORTATION

An important trend in transportation is the development of lightweight materials for automobiles that increase energy efficiency while reducing emissions. Here the key issue is the strength-to-weight ratio versus cost. Advanced composites with polymer, metal, or ceramic matrix and ceramic reinforcement are already in use in space systems and aircraft. These composites are too expensive for automobile applications, so aluminum alloys are being developed and introduced in cars such as the Honda Insight, the Audi A8 and AL2, and the GM EV1. Although innovation in both design and manufacturing is needed before such all-aluminum structures can become widespread, aluminum content in luxury cars and light trucks has increased in recent years. Polymer matrix, carbon-fibre (C-fibre) reinforced composites could enable high mileage cars, but C-fibre is currently several times more expensive than steel. Research sponsored by the Department of Energy (DOE) at Oak Ridge National Laboratory is working to develop cheap C-fibres which could have wider application and effect. Spurred by California's regulations concerning ultra-low-emission vehicles, both Honda and Toyota have introduced gasoline-electric hybrid vehicles.

The U.S. government and industry consortium called Partnership for a New Generation of Vehicles (PNGV) has demonstrated prototype hybrid vehicles that use both diesel/electric and diesel/fuel-cell power plants and has established 2008 as the goal for a production vehicle. These vehicles use currently available materials, but the cost reduction issues described will be critical in bringing production costs to levels that will allow significant market penetration.

ENERGY SYSTEMS

If the ready availability of oil continues, it may be difficult for technology trends to be much of a driving force in global energy between now and 2015. Key questions have to do with continued oil imports, continued use of coal, sources of natural gas, and the fate of nuclear power. Nevertheless, technology may have significant effects in some areas. Along with investments in solar energy, current investments in battery technology and fuel cells could enable continued trends in more portable devices and systems while extending operating times. Developments in materials science and engineering may enable the energy systems of 2015 to be more distributed with a greater capability for energy storage, as well as energy system command, control, and communication. High-temperature superconducting cables, transformers, and storage devices could begin to increase energy transmission and distribution capabilities and power quality in this time frame. The continued development of renewable energy could be enhanced by the combination of cheap, lightweight, recyclable materials (and perhaps the genetic engineering of biomass fuels) to provide cost-effective energy for developing countries without existing, well-developed energy infrastructures as well as for remote locations. Significant changes in developed countries, however, may be driven more by existing social, political, and business forces, since the fuel mix of 2015 will still be strongly based on fossil fuels.

Environmental concerns such as global warming and pollution might shift this direction, but it would likely require long-term economic problems (e.g., a prolonged rise in the price of oil) or distribution problems (e.g., supplies interrupted by military conflicts) to drive advances in renewable energy development.

NEW MATERIALS

Materials research may provide improvements in properties by 2015 in a number of additional areas, leading to significant effects. SiC, GaN, and other wide band gap semiconductors are being investigated as materials for high-power electronics. Functionally graded materials (i.e., materials whose properties change gradually from one end to the other) can form useful interlayers between mechanically, thermally, or electrically diverse components. Anodes, cathodes, and electrolytes with higher capacity and longer lifetime are being developed for improved batteries and fuel cells. High-temperature (ceramic) superconductors discovered in 1986 can currently operate at liquid nitrogen (rather than liquid helium) temperatures. Prototype devices such as electrical transmission cables, transformers, storage devices, motors, and fault current limiters have now been built and demonstrated. Niche application on electric utility systems should begin by 2015 (e.g., replacement of underground cables in cities and replacement of older substation transformers). Nonlinear optical materials such as doped LiNbO3 are being investigated for ultraviolet lasers (e.g., to enable finer lithography). Efforts are under way to increase damage threshold and conversion efficiency, minimize divergence, and tailor the absorption edge.

Hard materials such as nanocrystalline coatings and diamonds are being developed for applications such as computer disk drives and drill bits for oil and gas exploration, respectively. High-temperature materials such as ductile intermetallics and ceramic matrix composites are being developed for aerospace applications and for high-efficiency energy and petrochemical conversion systems.

NANOMATERIALS

This area combines nanotechnology and many applications of nanostructured materials. One important research area is the formation of semiconductor "quantum dots" (i.e., several nanometer-size, faceted crystals) by injecting precursor materials conventionally used for chemical-vapour deposition of semiconductors into a hot liquid surfactant. This "quantum dot" is in reality a macromolecule because it is coated with a monolayer of the surfactant, preventing agglomeration. These materials photoluminesce at different frequencies (colours) depending upon their size, allowing optical multiplexing in biological labeling. Another important class of nanomaterials is nanotubes (the open cylindrical sisters of fullerenes). Possible applications are field-emission displays, nanoscale wires for batteries, storage of Li or H_2, and thermal management (heat pipes or insulation—the latter taking advantage of the anisotropy of thermal conductivity along and perpendicular to the tube axis).

Another possibility is to use nanotubes (or fibres built from them) as reinforcement for composite materials. Presumably because of the nature of the bonding, it is predicted that nanotube-based material could be 50 to 100 times stronger than steel at one-sixth of the weight if current technical barriers can be overcome. Nanoscale structures with desirable mechanical and other properties may also be obtained through processing. Examples include strengthening of alloys with nanoscale grain structure, increased ductility of metals with multi-phase nanoscale microstructure, and increased flame retardancy of plastic nanocomposites.

NANOTECHNOLOGY

Much has been made of the trend toward producing devices with ever-decreasing scale. Many people have projected that nanometer-scale devices will continue this trend, bringing it to unprecedented levels. This includes scale reduction not only in microelectronics but also in fields such as MEMS and quantum-switch-based computing in the shorter term. These advances have the potential to change the way we engineer our environment, construct and control systems, and interact in society.

Nanofabricated Computation Devices

Nanofabricated Chips.

SEMATECH – the leading industry group in the semiconductor manufacturing business – is calling for the development of nanoscale semiconductors in their latest International Technology Roadmap for Semiconductors (ITRS). The roadmap calls for a 35 nm gate length in 2015 with a total number of functions in high-volume production microprocessors of around 4.3 billion. For low-volume, high-performance processors, the number of functions may approach 20 billion. Corresponding memory chips (DRAMs) are targeted to hold around 64 gigabytes. These roadmap targets would continue the exponential trend in processing power, fueling advances in information technology. Although a number of engineering challenges exist (such as lithography, interconnects, and defect management), obstacles to achieve at least this level of performance do not seem insurmountable. Given unforeseen shortfalls in the economic production of these chips (e.g., because of very high manufacturing costs or unacceptably large numbers of manufacturing defects), several alternatives seem possible. Defect-tolerant computer architectures such as those prototyped on a small scale by Hewlett-Packard offer one alternative. These alternative methods provide some level of additional robustness to the performance goals set by the ITRS. However, in the years following 2015, additional difficulties will likely be encountered, some of which may pose serious challenges to traditional semiconductor manufacturing techniques.

In particular, limits to the degree that interconnections or "wires" between transistors may be scaled could in turn limit the effective computation speed of devices because of materials properties and compatibility, despite incremental present-day advances in these areas. Thermal dissipation in chips with extremely high device densities will also pose a serious challenge. This issue is not so much a fundamental limitation as it is an economic consideration, in that heat dissipation mechanisms and cooling technology may be required that add to total system cost, thereby adversely affecting marginal cost per computational function for these devices.

Quantum-Switch-Based Computing.

One potential long-term solution for overcoming obstacles to increased computational power is computing based on devices that take advantage of various quantum effects. The core innovation in this work is the use of quantum effects, such as spin polarization of electrons, to determine the state of individual switches. This is in contrast to more traditional microelectronics, which are based on macroscopic properties of large numbers of electrons, taking advantage of materials properties of semiconductors. Various concepts of quantum computers are attractive because of their massive parallelism in computation, but they are not anticipated to have significant effects by 2015. These concepts are qualitatively different from those employed in traditional computers and will hence require new computer architectures. The types of computations (and hence applications) that can be quickly performed using these computers are not the same as those readily addressed by today's digital computer. Several workers in the area have devised algorithms for problems that are very computationally intensive (and thus time-consuming) for existing digital computers, which could be made much faster using the physics of quantum computers.

Examples of these problems include factoring large numbers (essential for cryptographic applications), searching large databases, pattern matching, and simulation of molecular and quantum phenomena. A preliminary survey of work in this area indicates that quantum switches are unlikely to overcome major technical obstacles, such as error correction, de-coherence and signal input/output, within the next 15 years. If this were indeed the case, quantum-switch-based computing does not appear to be competitive with traditional digital electronic computers within the 2015 timeframe.

Bio-Molecular Devices and Molecular Electronics

Many of the same manufacturing and architectural challenges discussed regarding quantum computing also hold true for molecular electronics. Molecular electronic devices could operate as logic switches through chemical means, using synthesized organic compounds. These devices can be assembled chemically in large numbers and organized to form a computer. The main advantage of this approach is significantly lower power consumption by individual devices. Several approaches for such devices have been devised, and experiments have shown evidence of switching behaviour for individual devices. Several research groups have proposed interconnection between devices using carbon nanotubes, which provide high conductivity using single molecular strands of carbon. Progress has been made toward raising the operating temperature of these switches to nearly room temperature, making the switching process reversible, and increasing the overall amount of current that can be switched using these devices. Several major outstanding issues remain with respect to molecular electronics. One issue is that molecular memories must be able to maintain their state, just as in a digital electronic computer. Also, given that the manufacturing and assembly process for these devices will lead to device defects, a defect-tolerant computer architecture needs to be developed.

Fabricating reliable interconnects between devices using carbon nanotubes (or some other technology) is an additional challenge. A significant amount of work is ongoing in each of these areas. Even though experimental progress to date in this area has been substantial, it seems unlikely (as with quantum computing) that molecular computers could be developed within the next 15 years that would be relatively attractive (from a price/performance standpoint) compared with conventional electronic computers.

BROADER ISSUES AND IMPLICATIONS

Examining the potential for developing qualitatively different computational capabilities from different technology bases is a challenging exercise. The history of computing over the last 50 years has seen one major shift in technology base (from vacuum tubes to semiconductor transistors), with a corresponding shift not just in computational power but also in attitudes about the value of computers. Ideas of computers as simple machines for computation gave way to the use of computers for personal productivity with the advent of the microprocessor. As the power of these microprocessors has grown exponentially, they have also been seen more recently as a vehicle for new media and socialization. The ramifications of future computing technologies will be determined principally by two factors: the conception, development, and adoption of new applications that require significantly more computational power; and the ability of technology to address these demands. New applications are always difficult to anticipate, but it is less challenging to foresee the likely consequences for diffusion of this technology.

Past experience with personal computers and telecommunications has shown that these technologies diffuse more rapidly in the developed world than in the developing world. It is difficult to foresee an increase in the political or ethical barriers to computing technologies beyond those seen today, and these are rapidly vanishing. On the remaining question of technology development, the odds-on favourite for the next 15 years remains traditional digital electronic computers based on semiconductor technology. Given the virtual certainty of continued progress in this area, it is hard to imagine a scenario in which a competing technology (quantum-switch-based computing, molecular computers, or something else) could offer a significant performance advantage at a competitive price. But the longer-term, traditionally elusive question in the period after 2015 is: How long will traditional silicon computing last? And when, if ever, will a competing technology become available and attractive? If an alternative computing technology becomes sufficiently attractive, the economic effects of technology substitution on the current semiconductor industry and adjacent industries must be considered.

Major industry players may be faced with a choice between cannibalizing their existing market opportunities in favour of these new, future technologies, competing head-on with new players, or simply acquiring them. Most important, given the very different architectural approaches of these technologies and the classes of problems for which they are best suited, what will be the effect on future applications? The promises of nanotechnology may indeed become a reality in the period after 2015, but it will face these competitive challenges before its significance becomes global.

INTEGRATED MICROSYSTEMS AND MEMS

MEMS is less an application area in itself than a manufacturing or fabrication technique that enables other application areas. Many authors use MEMS as shorthand to imply a number of particular application areas. As it is used here, MEMS is a "topdown" fabrication technology that is especially useful for integrating mechanical and electrical systems together on the same chip. It is grouped in the category of integrated microsystems because these same MEMS techniques can be extended in the future to also help integrate biological and chemical components on the same chip, as discussed below. Thus far, MEMS techniques have been used to make some functional commercial devices such as sensors and single-chip measurement devices. Many researchers have used MEMS technologies as analytical tools in other areas of nanotechnology such as the ones discussed here.

SMART SYSTEMS-ON-A-CHIP (AND INTEGRATION OF OPTICAL AND ELECTRONIC COMPONENTS)

Simple electro-optical and chemical sensor components have already been successfully integrated onto logic and memory chip designs in research and development labs. Likewise, radio frequency component integration in wireless devices is already being produced in•mass quantities. Some companies have products capable of doing elementary DNA testing. The 1999 ITRS predicts the introduction of chemical sensor components with logic in commercial designs by 2002, with electro-optical component integration by 2004, and biological systems integration by 2006. Given these predictions, there is clearly time for relatively complex integrated systems and applications to develop within the 2015 timeline. These advances could enable many applications where increased integrated functionality can become ubiquitous as a result of lower costs and micro-packaging.

MICRO/NANOSCALE INSTRUMENTATION AND MEASUREMENT TECHNOLOGY

Instrumentation and measurement technologies are some of the most promising areas for near-term advancements and enabling effects. As optical, fluidic, chemical, and biological components can be integrated with electronic logic and memory components on the same chip at marginal cost, drug discovery, genetics research, chemical assays, and chemical synthesis are all likely to be substantially affected by these advances by 2015. Some of the first applications of nanoscale (and microscale) instruments were as basic sensors for acceleration (such as those used in airbags), pressure, etc. Small, microscale, special-purpose optical and chemical sensors have been used for some time in sophisticated laboratory equipment, along with microprocessors for signal processing and computation.

Already, companies have produced products that allow for basic DNA analysis, and that can assist in drug discovery. As these sensors become more sophisticated and more integrated with computational capability (with the aid of systems-on-a-chip), their utility should grow tremendously, especially in the biomedical arena.

BROADER ISSUES AND IMPLICATIONS

There are several advantages of nanotechnology for integrated systems in general (and instrumentation and measurement systems as a subset of these). First, existing semiconductor technologies will likely allow the volume manufacture of integrated smart systems that can be produced at low enough cost to be considered disposable. Second, the massive parallelism afforded by this same technology allows for the rapid analysis (with integrated computation) of very complex samples (such as DNA), the processing of large numbers of samples, and the recognition of large numbers of agents (e.g., infectious agents and toxins). Devices with these properties are already of tremendous utility in the biomedical arena for drug testing, chemical assays, etc. In addition, they will likely find utility in a variety of industrial applications. Integrated micro/nanosystems are already starting to affect applications where miniaturization of components, subsystems, and even complete systems is significantly reducing device size, power, and consumables while introducing new capabilities. This area lends itself naturally to the confluence of all the broad areas discussed in this report (biotechnology, materials, and nanotechnology). The next five to ten years will likely see the integration of computational capabilities with biological, chemical, and optical components in systems-on-a-chip. At the same time, advances in biotechnology should drive applications for drug discovery and genomics, as well as the basic understanding of many other phenomena.

Advances in biomaterials will likely produce biologically compatible packaging, capable of isolating substances from the body in a time-controlled fashion (e.g., for drug delivery). The confluence of these capabilities could allow for continued development of microscale and nanoscale systems that could continue to be introduced into the body to perform basic diagnostic functions in a minimally invasive way, providing new abilities to remedy health problems. Other possible applications include: pervasive, self-moving sensor systems; nanoscrubbers and nanocatalysts; even inexpensive, networked "nanosatellites." So-called "nanosatellites" are targeting order-of-magnitude reductions in both size and mass (e.g., down to 10 kg) by reducing major system components using integrated microsystems. If successful, this could economize current missions and approaches (e.g., communication, remote sensing, global positioning, and scientific study) while enabling new missions (e.g., military tactical space support and logistics, distributed sparse aperture radar, and new scientific studies). In addition, advances could empower the proliferation of currently controlled processing capabilities (e.g., nuclear isotope separation) with associated threats to international security. Progress will likely depend on investment levels as well as continued S and T development and progress.

MOLECULAR MANUFACTURING AND NANOROBOTS TECHNOLOGY

A number of experts have put forth the concept of molecular manufacturing where objects are assembled atom by atom (or molecule by molecule). Bottom-up molecular manufacturing differs from microtechnology and MEMS in that the latter employ top-down approaches using bulk materials using macroscopic fabrication techniques. To realise molecular manufacturing, a number of technical accomplishments are necessary. First, suitable molecular building blocks must be found. These building blocks must be physically durable, chemically stable, easily manipulated, and (to a certain extent) functionally versatile. Several workers in the field have suggested the use of carbon-based diamond-like structures as building blocks for nano-mechanical devices, such as gears, pivots, and rotors. Other molecules could also be used to build structures, and to provide other integrated capabilities, such as chemically reactive structures. Much additional work in the area of modeling and synthesis of appropriate molecular structures is needed, and a number of groups are working to this end. Dresselhaus and others are fabricating suitable molecular building blocks for these structures. The second major area for development is in the ability to assemble complex structures based on a particular design. A number of researchers have been working on different approaches to this issue. Different techniques for physical placements are under development. One approach by Quate, MacDonald, and Eigler uses atomicforce or molecular microscopes with very small nanoprobes to move atoms or molecules around with the aid of physical or chemical forces. An alternative ap- proach by Prentiss uses lasers to place molecules in a desired location.

Chemical assembly techniques are being addressed by a number of groups, including Whiteside's approach to building structures one molecular layer at a time. A third major area for development within molecular manufacturing is systems design and engineering. Extremely complex molecular systems at the macro scale will require substantial subsystem design, overall system design, and systems integration, much like complex manufactured systems of the present day. Although the design issues are likely to be largely separable at a subsysiems level, the amount of computation required for design and validation is likely to be quite substantial. Performing checks on engineering constraints, such as defect tolerance, physical integrity, and chemical stability, will be required as well. Some workers in the area have outlined a potential path for the evolution of molecular manufacturing capability, which is broken down by overall size, type of fabrication technology, system complexity, component materials used, etc. Some versions of this concept foresee the use of massively parallel nanorobots or scanning nanoprobes to assemble structures physically (with 100 to 10,000 molecular parts). Other, more advanced concepts incorporate chemical principles and use simple chemical feedstocks to achieve much larger devices on the order of 10^8 to 10^9 molecular parts. Ostensibly, as each of these techniques matures (or fails to develop), more systems and engineering-level work must be done before applications can be realised on a significant scale.

Although molecular manufacturing holds the promise of significant global changes (such as retraining large numbers of manufacturing workforces, opportunities for new regions to vie for dominance in a new manufacturing paradigm, or a shift to countries that do not have legacy manufacturing infrastructures), it remains the least concrete of the technologies discussed here. Significant progress has been made, however, in the development of component technologies within the first regime of molecular manufacturing, where objects might be constructed from simple molecules and manufactured in a short amount of time via parallel atomic force microprobes or from simple self-assembled structures. Although the building blocks for these systems currently exist only in isolation at the research stage, it is certainly reasonable to expect that an integrated capability could be developed over the next 15 years. Such a system could be able to assemble structures with between 100 and 10,000 components and total dimensions of perhaps tens of microns. A series of important breakthroughs could certainly cause progress in this area to develop much more rapidly, but it seems very unlikely that macro-scale objects could be constructed using molecular manufacturing within the 2015 timeframe.

BROADER ISSUES AND IMPLICATIONS

The present period in molecular manufacturing research is extremely exciting for a number of reasons. First, many workers have begun to experimentally demonstrate basic capabilities in each of the core areas outlined. Second, continued progress and ongoing challenges in the area of top-down microelectronics manufacturing are pushing existing capabilities closer to the nanoscale regime. Third, the understanding of fundamental properties of structures at the nanoscale has been greatly enhanced by the ability to fabricate very small test objects, analyse them experimentally using new capabilities, and understand them more fundamentally with the aid of sophisticated computational models. At the same time, many visionaries have advanced notions about potential applications for molecular manufacturing. But because experimental capabilities are in their infancy (as many workers have pointed out), it is extremely difficult to foresee many outcomes, let alone assess their likelihood. International competition for dominance or even capability in cutting-edge nanotechnology may still remain strong, but current investments and direction indicate that the United States and Europe may retain leadership in most of this field.

Progress in nanotechnology will depend heavily on R and D investments; countries that continue to invest in nanotechnology today may lead the field in 2015. In 1997, annual global investments in nanotechnology were as follows: Japan at $120 million, the United States at $116 million, Western Europe at $128 million, and all other countries (former Soviet Union, China, Canada, Australia, Korea, Taiwan, and Singapore) at $70 million combined. Funding under the U.S. National Nanotechnology Initiative is proposed to increase to $270 million and $495 million in 2000 and 2001, respectively.

This would not preclude other countries from acquiring capabilities in nanotechnology or in using these capabilities for narrow technological surprise or military means. Given the difficulty in foreseeing outcomes and estimating likelihoods, however, it is also difficult to extrapolate predictions of specific threats and risks from current trends.

5

Cultivating Waste

EXCESSIVE RISKS IN AN ECONOMY OF OPPORTUNITIES

In this chapter we shall deploy the triad of risk society thesis, Actor Network Theory and biophilosophy to theorize waste. The purpose of theorizing waste is to show how what seems to be merely a 'side-effect' of the capitalist political economy not only poses serious problems for the sustainability of life in a capitalist world order, but—in a paradoxical moment—also transforms the very nature of that order into what Beck referred to as 'the world risk society'. As waste becomes big business, in terms of both financial and symbolic capital, the logic of 'goods' and its associated disciplinary regime become subverted. This creates an ambivalence in which waste itself accumulates, both in terms of valorization and pathogenicity. The chapter starts with a brief assessment of the relationship between waste and risk, arguing that the inherently excessive nature of risk makes its 'objectification' (in science and engineering, as well as commerce and governance) an incredibly volatile process.

We then look at how this volatility affects the way in which waste has been politicized in the age of industrial modernity, before focusing on the 'turning' towards what shall be referred to as 'the biopolitics of waste'. This 'turning' has affected the way in which technological culture has responded to the risk challenges revealed by technoscience, and has specifically evolved around the commodification of waste. That waste is not just big business is subsequently illustrated by drawing on Nigel Clark's work on 'nanotechnology'. Using his critique of some interpretations of the risk society thesis, the chapter concludes with a reconceptualization of 'waste-risk' as a virulent abject of modernity, poised to inaugurate a pathogenicity that is at once ephemeral yet deeply disturbing.

WASTE AND RISK

Waste is perhaps the most universal, vulgar and banal example of ecological risks in everyday life. Nearly all ecological risks relate in one way or another to waste, and more specifically pollution. Whether nuclear, genetic or biochemical, ecological risks are materially engendered by pollutant waste objects—atoms, genes, molecules and cells. However, on a more mundane level, waste is less differentiated (even if separated) and appears as one amorphous blob of matter out of place. Waste is often linked to disease. It represents uncontrolled matter-out-of-place, freely interacting and reacting, cultivating bacteria, fungi and toxins that may pose direct threats to our health. Ever since the connections had been made between germ theory and the sanitation movements in western Europe and the USA in the late nineteenth century, the regulation and expulsion of waste from the domestic and public spheres has been a governmental priority, although not everywhere as Craddock's historical account of San Francisco's rather lax regime of public health management illustrates. In many ways even today, public health management still espouses the values of the nineteenth-century sanitarians. The introduction of closed sewage systems exemplifies the articulated need to separate waste flows from all other flows in the domestic and public domains.

The removal of waste-dumps from urban areas, and from areas of human inhabitation generally, equally signifies this need to separate waste from ordinary life. Beck's risk society thesis, strongly evolves around associations between industrial modernization, waste and tangible and intangible risks. Waste, produced as side-effects of industrial production, engenders latent risks to human health. A clear example of this is the relationship between waste and disease. First there is a direct link between waste and infectious disease. Typhoid fever, cholera and gastroenteritis are all caused by types of bacteria that live on (human) waste. When human waste is not removed from drinking water, such infectious and virulent bacteria can easily reproduce themselves. Cultivated in the sludge of human and animal excreta, water and offal, they pose an immediate threat to human health. Needless to say, their presence is most ubiquitous in 'open' sewage systems. Second, the relationship between waste and disease can be indirect. Here we should particularly focus on the role of animals and parasites such as rats, mice, fleas and insects. *Yersinia pestis,* or the Plague, for example, is carried by rat fleas. Rats and humans often cohabit, especially when there is a lot of waste around. Cockroaches and houseflies are equally prolific carriers of diseases and they too thrive on waste. It was against such kinds of risk associations with waste that the earliest public health reforms were initiated in the nineteenth century.

These initiatives, spearheaded by health reformers, hygiene movements and medical scientists, were the primary forces behind the widespread introduction of sewage systems, organized refuse collection, incineration and the use of landfill dumps outside areas of human habitation. However, not all waste-risks are that easily overcome. Especially risks related to the toxicity of certain chemicals of matter-out-of-place have proven to be more difficult to remove. Sometimes, however, these risks did become spectacularly manifest—as in the *Chernobyl, Bhopal, the Exxon Valdez and Torrey Canyon disasters.* In all four cases, the relationship between industrial production, waste and risk were revealed in the dramatic consequences on the environment. Furthermore, the fact that in these cases waste-risks were inseparable from the nature of 'goods' produced by the industrial production itself, served as a reminder that waste-risk is never simply an undesired side-effect of modernization, but endemic to it. More often

domains of modern life. As excess, waste thus engenders a volatility. Hovering between presence and absence, its ubiquitous ephemerality poses serious problems even for a culture based on strong boundaries of inclusion and exclusion. The excessive nature of waste returns to haunt the present as past-waste becomes present-toxicity. Suddenly almost all ecological risks could be rearticulated as pollution or waste-risks; indeed, one could argue that the world risk society is above all a world waste society.

6

Cyberrisks

TELEMATIC SYMBIOSIS AND COMPUTER VIRUSES

You know—technology wasn't invented by us humans. Rather the other way around. As anthropologists and biologists admit, even the simplest life forms, infusoria (tiny algae synthesized by light at the edges of tidepools a few million years ago) are already technical devices. Any material system is technological if it filters information useful to its survival, if it memorizes and processes that information and makes inferences based on the regulating effect of behaviours, that is, if it intervenes on and impacts its environment so as to assure its perpetuation at least. Viral epidemics present a case of abject virulence which has co-evolved with humans, animals and plants since the very beginning. Their existence as such is not specific to modernity's technological culture. However, we argued that 'emergent pathogen virulence' is a typical example of the 'turning' of technological culture, indeed perhaps even a messenger of the incumbent apocalypse. This is because the logic of emergence is intertwined with the limits of technological culture.

As it expands into previously unknown 'junctural zones', especially tropical rainforests, this technological culture challenges and reveals nature's darkest secrets. Moreover, the continued exploitation of life, not just in terms of capitalism but in terms of the very underpinning of the technical systems that 'run' this economy, including its submolecular level, has created 'hot zones' of abject virulence, marked by high levels of pollution, genomic interaction and infection. However, all these instances of abject virulence could be summarized by the subtitle of Geoffrey Cannon's book—*Nature's Revenge.* It is based on the idea that 'man's' asserted dominion over nature, through the deployment of technoscience, is now ending. Emergent viruses are merely the payback of decades of unbridled and careless commercial exploitation of nature, especially the rainforests. Nature has interesting ways of balancing itself. The rain forest has its own defences. The earth's immune system, so to speak, is seeing the presence of the human species and is starting to kick in. The earth is attempting to rid itself of an infection by the human parasite. This view, however, has its own limits. It assumes that nature itself is a coherent, homeostatic system (i.e. ecosystem) that, if not motivated by retribution, is motivated by a desire to restore its own order.

This view contradicts the evolutionist perspective which understands nature as a process of continuous transformation. Moreover, it sets out a distinction between humans and nature, as if the former are separate from the latter and thereby fails to acknowledge the ambivalence of nature, as itself following the logic of a technical system, be it that of homeostasis or evolution. Moreover, this view is problematic because it assumes 'man' to be merely a catalyst in the unfolding of otherwise completely natural risks. It forecloses any understanding of risks as part of an assemblage of continuous flows of differentiation and modulation. Finally, such a view is deeply misanthropic. It rests on a Malthusian domesday scenario which sees overpopulation as the source of all evil. The eugenic misanthropy is particularly directed at the Third World, where the problems are most severe, yet thereby negates the central role played by the affluent west, whose wealth is largely derived from worldwide exploitation. It thus endorses a highly questionable politics of 'blaming the victim'. Instead of promoting a misanthropic notion of nature's revenge, this chapter seeks to widen the 'turning' of technological culture in the risk society by looking at risks that can never be idealized as 'natural'.

The starting point is a problematization of 'nature' based on Macnaghten and Urry's work *Contested Natures,* which will be broadened into a problematization of 'the environment' as a whole. Appropriating the concept of autopoiesis, it will be made clear that a cybernetic conception of life as information processing will not only annihilate the unhelpful distinction between nature and culture, but also enable a conception of technological culture as entropie. Abject virulence is thus not a regulating device of systemic integrity, but a radically transformative and vital force in the reconstitution of life. The concept of 'cyberrisk' will be invoked to open up the scope of analysis to the material-informational technological processing of risks that comes before their discursive realization in signifying practices.

EXPOSING 'THE' ENVIRONMENT: A CYBERNETIC TURN

In his acclaimed *Rise of the Network Society,* Manuel Castells describes the contours of an informational landscape characterized by a series of relatively smooth, globally operational, flows (of data, people, money, goods, services, images). This landscape, or what he calls 'space of flows', is the field in which global capitalism (with its geopolitics of the new world order) is able to bolster new heights in the accumulation of wealth and power. The space of flows, however, is further marked by an increased separation from what he calls 'the space of places' – the lived geographies of everyday life. It is the growing gap between these two landscapes, the space of flows and the space of places, that is the definitive condition of inequality in the informational society. Irrespective of whether one agrees with Castells's vision of the future or not, his notion of an informational landscape is worthwhile pursuing. It highlights the problematic of conceptualizing 'the environment' in an age in which technoscience has irreversibly imposed its own specific logic onto everything else.The environment is itself the product of an internalized differentiation of specific systems. The environment is a label that is assigned to that which remains; it is the by-product of the technological enframing of the world that has been called 'modernity'. In this sense, the term 'environment' shares the conceptual space reserved for 'waste', 'dirt', 'residue', 'excess' and 'risk'. It is thus intrinsically contingent upon those systems which order the world: technoscience, governance (including politics and law), media and commerce. Throughout this book it has been argued that technology reveals and orders the world. It produces a specific relationship between the existential being calls *Dasein*) and its being-in-the-world. This 'in-the-world' could be conceptualized as basically an 'environment of being', a space in which being emerges. It has also been argued that technological culture orders and reveals a specific relationship between human being and 'his' environment. This relationship is principally enframed in terms of a challenging-revealing of the environment in support of 'man's' full self-realization as an autonomous being. That is, technology offers itself in service of 'man' and constitutes a relationship between 'man' and 'his' environment in terms of what it orders and reveals. However, technological culture does more. It enables 'man' to encounter 'his' environment carelessly. That is to say, it engenders a 'forgetting' of concern for being outside the technical system which mediates the encounter. Technological culture generates enclosure through mediation by ordering an environment. This is clearly shown by Macnaghten and Urry's critical analysis of the way in which different conceptions of 'Nature' are actualized in social practices. The different conceptions of 'Nature' simultaneously posit 'it' as a singular, integral and self-evident entity.

That is, whereas there have always been competing constructions of Nature, the incompatibilities have only relatively recently become subject to more overt political struggle. Indeed, even if the dominant conception of 'Nature' as an externalized environment still prevails, the contestations of such a conception have now permeated all aspects of technological culture. The concept of 'virtual object' has been used to argue that the process of 'enpresenting' involves a complex chain of revelations and orderings that engender a multiplicity of actualizations. We distinguished between presentation, representation and re-presentation to show that technoscience is merely one way of 'making present' and albeit the dominant one in technological culture, it cannot foreclose alternatives. Technoscience is mostly concerned with representation and as a result operates on virtual objects *as if* they are real. Until recently, scientific realism was largely hegemonic and hardly contested, at least outside the humanities and social sciences. Environmental risks are simply conceived as hazards with probability calculations attached to them. This, however, can no longer be taken for granted. Whereas there have always been sound philosophical arguments against the naïve realism of risk assessments, this chapter argues that there is now an overwhelming historical-realist case against it as well.

This argument, perhaps originally posed by Lyotard in his playful, perhaps even misleading, concept of postmodernism, was further developed by theorists such as Baudrillard, Jameson and Vattimo to stress that the very same technologies of representation which inaugurated technoscience as the superior mode of revealing, have now turned against the very foundations of this power. Especially, with the advent of digital information and communication technologies any stable and fixed notion of 'reality' has been undermined. The Gulf War gave us images of 'smart' bombing that may actually have been produced by a simulator. Images of Palestinians dancing in the street after the news of the bombing of the World Trade Centre on 11 September may have been more than ten years old. We live in a hyperreality. Our environment is now so heavily mediated that we cannot even trust our own empirical senses to give us direct access to the real. This 'real' used to be the environment of our being in the world. However, the environment of ICTs is always-already subject to continuous virtualization. Digitalization, especially, is a force that shifts power away from atoms to bits.

It is telling that genetics is modelled on information science, not the other way round; this dates back to the early 1950s. Our very concept of the logic of life is modelled on technology. This is not new. The Greek word *organon,* which serves as the basis for organism as well as organization, originally referred to machinery. For the world of ICTs, the environment is always a virtuality—which can be realised only through deferral. The space of flows of the informational landscape discloses 'an' environment that is endlessly deferred in its enframed 'out-there-ness'. The contingent singularity of this endless deferral, or differance, is nothing more than the actualization of a particular technology, or medium. ICTs have also enabled a new kind of sensibility within technological culture. This sensibility is marked by a rapid acceleration of information flows: the future becomes increasingly 'present' as an extended presence.

This has created a possibly new kind of episteme which could be called 'hyperthought'. The difference between modern thought and hyperthought is that whereas in the first episteme environment remained contingent upon integral or molar media-forms (science, governance, media, commerce), in hyperthought environment is enacted by a perpetual differential or modular assemblage of mediaforms (science-becoming-governance-becoming-media-becoming-commerce). Indeed, it is the modulation of the different flows that spatialize the informa- tional landscape into particular media assemblages that characterizes this new hyper-contingent sense of 'environment'. The deadlock between reality and representation, referent and sign, objectivity and subjectivity that constitutes the episteme of modern thought is suddenly broken with the advent of new information and communication technologies. Indeed, the world has changed in many ways. Of course, we still have capitalism, we still have nation-states, parliamentary democracy, disciplinary power, racism and patriarchy, but we must seriously question whether any of these labels still 'signify' the same 'phenomena' and inevitably, whether the very labels 'signification' and 'phenomena' are what they used to be. We should not be misled in thinking that everything should be different in order to find a rupture or discontinuity.

That is a typically modernist presupposition ('a rupture is when everything changes'). Perhaps the difficulty we face now is that change is not as sweeping, clear-cut, totalitarian and all-pervasive as 'the Great Transformation' (Industrial plus French Revolutions). It seems as if we are in the midst of a perpetual rupturing as new worlds ('the' informational society being one of them) are hailed and forgotten by the flick of a page (or screen). Modernity is valorized by 'the' new but in a world that has lost the unity of perspective, 'the' loses its special particular-universalist character. Speed is so intense that the very possibility of sustained newness is vanishing rapidly. When considering an environment such as that of new information and communication technologies, we cannot but account for its technological actualization. We may project such an environment to be everything 'outside' of the technological enframing of information processing and digital communication, it could even encompass the entire planetary physical world, including the media worlds of books, cinema, television and radio; however, this is possible only if we also acknowledge that which is 'inside', or more accurately, that which produces the difference between inside and outside, i.e. *autopoiesis*. Social theoretical conceptions of autopoiesis often operate within a scientific paradigm founded by system theorists such as Karl Deutsch and Norbert Wiener. The latter devised a term for it: cybernetics. It is this paradigm that may be seen as fundamental to the development of artificial intelligence (AI) in the 1970s and the subsequent information-processing and telecommunications revolutions of the 1980s. Cybernetics concerns the organization of the processing flow of information and the ability not only to reproduce itself, but also to adapt to and modify its environment in order to secure sustainability. Quite clearly, autopoiesis is a fundamental process in cybernetics—it is the becoming-self-reliance of an information-processing device. However, cybernetics also highlights something that Luhmann and Teubner seem to ignore: namely *symbiosis*.

For the theorists of autopoiesis, life is a process of chance—a coincidence established on the basis of the unlikely autopoietic take-off, when an information-processing device becomes an organism. However, studying viruses, it is quite obvious that life is not exclusively limited to autopoietic forms. The life of viruses is dependent upon its capacity to live alongside its host, that is, to establish a symbiotic relationship. It is symbiosis that allows systems to induce each other, to transform their information-processing flows. This is the core of what we are experiencing as the telematic revolution. Telematics refers to the interconnection between systems of telecommunications and information processing, allowing for instantaneous, worldwide dissemination of data. It operates on the basis of a very simple principle of digitalization. Whatever can be transformed and reproduced in digital form can be processed via telematics—text, sound, image, even neural and tactile-sensory experiences. It is thus not so much the arrival of new media, but the *assemblage* of media technologies, that marks the telematic revolution. The assemblage combines simplicity with speed.

It allowed for new powerful intersystemic symbioses as media technologies have started to pervade the military-industrial complex, large corporations, the state apparatuses, the culture industries and increasingly our private homes. Satellite, telephone, cable, computer, television operate together; the information flow allows for the transcoding of all kinds of other flows: finance, signs, goods, services, even people. The speed with which the telematic symbiosis has established itself is phenomenal; and there is no turning back. The transcoding is infinite as nanotechnologies rewrite DNA—'nature' becomes just another series of ones and zeros. 'The' environment, which was once discursively protected by a clear-cut and intransgressible distinction between nature and culture, is now up for grabs—open to any form of decoding and transcoding. Consequently, we would be quite mistaken to separate an environmentalist discourse about the natural world from discourses of telematic symbiosis. In the midst of the speed and irreversibility, 'man'—who already lost a centred consciousness in twentieth-century philosophy and psychoanalysis—is now also rapidly losing the plot in managing information flows. Reading has become browsing, reflecting now takes place by the move of a cursor on a hyperlink and the mere click of a mouse.

Telematic symbiosis has surpassed our ability to master and control it and subject it to our programming of the future; instead our futures are being programmed as we speak. This sense of loss manifests itself in the uncanny but perpetual return of the same debate over the merits and evils of cybertechnologies. On the one hand, the technophiles and technocrats cannot stop exaggerating on all the goods of the virtual: we are no longer hampered by the inertia of the flesh, construct our identities as we please, are free to engage in whatever we like and become participants in a more open and equal form of civic society. On the other hand, the technophobes and neo-luddites do not stop complaining about the dumbing and numbing effects of telematics. Telematics as nothing but an extension of capitalist exploitation, politics of propaganda and mystification of western imperialism, turns us into apathetic, lethargic, brainwashed consumers of infotainment. Both extremes show a remarkable inability to sensitize the inherent ambiva- lence of the telematic symbiosis and operate as if in a permanent neurotic narcosis. Both extremes are still trapped in a technological culture, governed by instrumental fidelity and humanist or its mirror image: misanthropic delusions.

They are not different from the arguments for and against technological innovations in the nineteenth century. As we have learned, technoscience continued to march on and revolutionize the world. However, there are grounds for a (tempered) optimism as well. Unlike in the nineteenth century, technoscience is no longer unified by a single episteme (spearheaded by positivism). Cracks and fissures have appeared and have induced more critical sensibilities. Risk has now revealed limits to technological culture that cannot be brushed aside by a mere reference to the inevitability of progress. It is here that the connection between telecommunication and data-processing has been most effective. Whilst being generated by a desire for speed and overcoming obstacles of time and space, telematics have also engendered new forms of alienation. Whilst initially such alienation could be seen as a primary force for identity-politics and projects of the self, an increasing number of people have given up attempts to unify themselves or the relationship between their bodies and subjectivities. This loosening of univalence between the body and subjectivity enables human beings to live their lives as decentred creatures. There is no need to read Derrida or Freud to understand the decentring of the subject; it has become common sense. It can be argued that the decentring of the subject finds its actualization not in philosophy but in telematics; it is here that the subject can finally come to practical terms with the externalization of her/his nervous system which—according to McLuhan—took place with the arrival of television.

A direct consequence of this, it can now be argued, is that the relationship between human beings and their environment is changing. The notion of 'man' as centre of the cosmos now looks as ridiculous as the pre-Copernican claim that the earth was the flat centre of the universe. It is less clear, however, what or who is to take 'his' place. The space of flows, or cyberspace, is not only a particular type of environment, but also one which problematizes the 'naturalistic' notion of 'the' environment and the implicit centredness of human consciousness therein. Environments are produced remainders of systemic self-referential differentiation (autopoiesis) and intersystemic symbiosis. They are formed through a process of territorialization, in which an assemblage creates a milieu, by recoding elements from other assemblages, that is, by deterritorializing other environments. The telematic revolution of the 1980s has exposed the inherent flaws in conceptualizing 'the' environment as an ecological 'natural' system. It is the transcoding of flows between the systems of science, governance, commerce, media, law and the military that marks the current climate of conceptualizing environmental risk. Whereas these systems may continue to operate on the basis of their own distinctive self-referential technologies and their inherent logic of, secrecy, expertise commodification or mediation, it has become obvious that the real power is exercised from the assemblage that emerges in-between/

beyond them—in the symbiosis of technologies of visualization, signification and valorization. Only when these forces are adding up, will a virtuality become 'realised'. Such a symbiosis is thus a closure of entropy, a gathering and ordering of multiplicities into a singular assemblage. In technological culture, this singularity was achieved through the domination of technoscientific codes of representation. Technoscience revealed and ordered the real.

In the risk society, this old symbiosis of representation is eroding and has started to turn against itself. The newly emergent telematic symbiosis thus entails a deterritorialization of the previously closed milieu of technological culture. A key feature of the assemblage is that both system and environment constitute each other interactively. Hence, it follows that if we argue that telematic symbiosis engenders a new multiplicity of (contested) environments, then this itself impacts upon the 'media systems' which have engendered the emergent symbiosis in the first place. Indeed, as with 'the' environment, we are witnessing some radical changes in the systemic organization of the 'mediascape'. We are witnessing a rapid hybridization of mediaforms which have sent panic waves of ambivalence throughout modern western societies, undermining the fixed boundaries and classification systems through which discipline and capital could organize and valorize a particular social, cultural, political and economic order. What characterizes this mediascape is a growing uncertainty over the boundaries between on the one hand the real and the possible, and on the other hand the virtual and the actual.

The telematic symbiosis of the mediascape has affected all 'domains' of disciplinary society: science, governance, law, the military, commerce, and on all levels from corporations and institutions to individual bodies and even genes. The blurring of the boundaries between the real and the possible has generated a crisis over realization. Increasingly, government agencies and corporate bodies are operating on the basis of complete uncertainty whether something can be realised or not. As the intervals of action are shrinking, time is vanishing. The speed of anticipation starts to overtake that of actualization. Technoscience, with its carefully cultivated and protected enclosures of expertise, is simply not up to the task. Governments cannot wait for laboratory experiments to confirm with 'scientific certainty' that something is a 'real' hazard or not. As definitions of what is hazardous are deeply political, a loss of 'technoscientific sanctimony' may actually be advantageous. It may open up debates about the perspective-bound relativity of risks and hazards. However, this optimism could also be premature, as new forms of subpolitics—judicial, military, commercial, media, administrative—are already waiting and eager to fill the void. Moreover, the blurring between the boundaries of the actual and the virtual has engendered a crisis of actualization, in which the distinctions between fact and fiction, or the referent and the signifier, have become increasingly void of any differentiating power.

The speed with which these boundaries are being blurred by the telematic symbiosis has caused a number of serious breakdowns in social organization. One only needs to look at politics or stock markets to understand the volatility of signification, as Peter Bernstein testifies:

Capital Markets have always been volatile, because they trade in nothing more than bets on the future, which is full of surprises. Buying shares of stock, which carry no maturity date, is a risky business. The only way investors can liquidate their equity positions is by selling their shares to one another: everyone is at the mercy of everyone else's expectations and buying power... Such an environment provides a perfect setting for non-rational behaviour: uncertainty is scary.

Coverage in this sense refers to a process of displacement: real environmental risks are replaced with signs; the signs are stand-ins for 'the real'. 'Critique' in this sense often refers to the adequacy of this 'standing in'—do the created signs actually correspond to that which they have replaced? Quite obviously, the crisis of actualization, which is inherent in the telematic symbiosis that is redesigning many mediascapes, obliterates any firm grounding of distinctions between the real and the representational, since both can only be expressions of a limit condition of the virtual. Fortunately, most critical analyses go beyond this and ask a much more productive question of what such media coverage *does.* Here we enter into a second axis of understanding representation, this time not as a process of replacement, but one of displacement. Representation is then not so much a confinement of the real in signification, but a specific realization of possibilities. Mediascapes are articulations of reality effects. Hence, the critique shifts from a question of correspondence to one of performativity—what is the coverage actually achieving? In media studies, answers are most often located in terms of 'covering-up'. Here, the media are seen as ideological state apparatuses that mystify environmental hazards and naturalize them.

Although there are clear reasons to argue that environmental hazards are often covered-up in mediascapes, as information about risks is often controlled by actors who may have a lot to lose, we do not need to place our sole faith in conspiracy theories. The point to make is that 'coverage' or representation does more—it is not only an act of concealment, but also one of unconcealment (in terms of sense-making). This requires a more productive (performative) understanding of mediascapes as particular technologies. Media representation operates on the basis of technologies of signification, thereby allowing other systems (including what we may call 'the public') to articulate and relate to perceived risks in particular ways—and thereby engaging in what may be called the (sub)politics of the risk society.

VIRTUAL RISKS AND TELEMATIC SYMBIOSIS

The question to be asked here is in what way can Beck's risk society thesis speak to an 'environment' that has already been completely incorporated by technologies of mediation? We have argued that it is in the very ambivalent role of technology/mediation—as both a making-sense (unconcealment) and a re-presentation (concealment) of risk—that the risk society reinstates the paradox of modern life as an order of disorganization. The power of Ulrich Beck's work on risk resides mostly in its visionary ability to see that we are witnessing a cultural-transformation of risk perception. This cultural transformation is predominantly, but not exclusively, organized around a particular rupture—the legitimation crisis of technoscience. The risk society calls upon the very same technoscience to both reveal and conceal the danger; the paradoxical consequence being that of technoscientific delegitimation and political indecisiveness—and a further eradication of the foundations of modern society. It is now common sense to associate technoscience with risk. Associated with both causes of and solutions to risk, science and technology have played an intricate part in the construction of economic, social, political and cultural conditions under which risks operate in modern society. Technoscience primarily operates through technologies of visualization—thereby generating particular perspectives and insights and constituting the condition for (mis)recognizing unspecified possibilities as 'risks'. Governance and mediascapes tap into that visualization with their respective technologies of signification: governance by articulating a distinction between 'significance' and 'insignificance' (hence difference and indifference), mediascapes by articulating sense and cultivating sensibility, bearing witness to the event of visualization and (governmental) signification as it unfolds.

It is often suggested that media invoke technoscientific sources to grant legitimacy to the coverage. The acquisition of such a legitimacy, however, is not the accomplishment of either science or mediascape, but of the assemblage between technologies of visualization and signification—the first by giving insight, the second by attributing meaning. To signify is to attribute meaning. The attribution of meaning is a form of encoding, of 'recovering' that which has been revealed. The stabilization of this recovery is the work of *logos*. *Logos* can be made to stand in for a range of other terms: language, law, logic, even reason or discourse. It always refers back to 'coverage' or 'concealment'. When something is put under the heading of *logos,* it is given a name, it becomes an attribute of the name, it becomes covered by the name. As Foucault pointed out, modern thought went even a bit further than merely 'naming' and 'classifying' and started to replace the name (language) with 'logic' (both in terms of causality and of law).

This can be traced in today's usage of the term ecology, which often implicitly refers to a carefully balanced 'natural' system in which the laws of nature work in a perfect symbiosis to establish harmony and homeostasis. Nowadays it is impossible to operate under the heading of ecology without submitting oneself to the technologies of visualization that characterize much of modern technoscience. Granting 'insight' as the basis of knowledge is the central focus. Even if there is a considerable and critical residue of knowledge outside of the technoscientific establishment, one cannot underestimate the powerful impact of the technoscientific overcoding—or overexposure, as Clark calls it—of such knowledge. This is because technoscience has been very effective in symbiotically appropriating the force of technologies of signification. Technoscience has successfully induced both governance and the mediascape to be able to intervene in the differentiation between significance and insignificance. This kind of 'subpolitics' does not engage with the same forms of accountability that characterize the political domain of representational democracy. What is nowadays called the public understanding of science can be seen as a programme of inducing the mediascape in order to more effectively operationalize the technologies of visualization. What happens to environmental hazards when they are overcoded by technoscience is that they are being transformed into objects of visual desire. The desire to make visible the implicit causal link between substance A, malign effect B, and possible effective treatment C (although the latter is much less clearly in the picture, it remains a promise) is a major motivation of contemporary scientific 'discovery'.

What happens as a result of this visualization, however, is the generation of more technoscientifically accurate 'knowledge' about hazards, which, via the enrolment of governance and mediascapes, are transformed into (sub)political issues for decision-making via signification and become risks. With 'the' environment already transcoded by the telematic symbiosis, and signification being induced by technoscientific desire, it is quite difficult to see any relevance in asking the question whether such environmental risks are real or constructions. They are always 'virtual'. What matters is the way in which these risks are being transformed: environmental risk becoming health risk becoming economic risk becoming political risk becoming representational risk. In a world of perfectly autopoietic systems (such as the disciplinary society) such risks would remain enclosed within one (sub)system only. With telematic symbiosis, however, they can move relatively smoothly across a range of intersystemic boundaries and at high speed. The result is a hyper-intensified multiplication of risks. Hence, environmental risks are no longer contained by any single particular system, be it that of science, governance or commerce. Telematic symbiosis has allowed risks to flow across systemic boundaries, inducing a range of environments and further contaminating and corrupting systemic integrity.

Indeed, far from being an event of technocratic supremacy, the assemblage that is currently being formed between technoscience, governance, media and commerce bears all the signs of a highly vulnerable and fragile stature. Hence, it seems perfectly logical to extend Ulrich Beck's analysis of risk society to the domain of electronic communications. Just as there have been applications to nuclear, genetic and reproductive technologies, so can the risk society thesis be applied to the world of information and communication technologies. Particular examples of such risks would be moral panics, such as outbreaks of 'irrational fear' after particular broadcasts, e.g. XTC-deaths, BSE, youth cultures, information overload and noise (e.g. middle management stress), cyberaddictions and cybercults even resulting in mass suicide (HeavensGate), hacking and cyberfraud (e.g. with credit card numbers). Finally, of course, the hallmark of cyberrisks is the computer virus, which represents, as Nigel puts it 'a new generation of digital demons to suddenly render our entire datasphere into terminal gibberish'. In order to understand the material-semiotic performativity of the mediation of environmental risk (that is, the question of how environmental risks operate when overcoded by a telematic symbiosis), we need to engage with the concept of cyberrisks.

CYBERRISKS

The simulation of nature is reaching this level of 'sophistication', raising the possibility of simulacra 'running wild', interacting, and even generating their own forms and functions...'irruptions' of simulated nature now appear to offer both the scientist and the general consumer of images the enchantment of a 'living' world which is constantly, rapidly, endlessly generating novelty. The risk society thesis works on the basis of the assumption that the technological culture inhabited by humans can no longer perform functional closures. That is, the self-evidentness of the ordering-revealing enabled by technoscience can no longer be taken for granted. As a result the relationship between human being and environment is no longer unproblematically mediated by technology; instrumental fidelity breaks down. This breakdown is catalysed by the digitalization of environments. In the world of telematics, the relationship between human being and environment is inevitably ambiguous. The environment is already multiplied by the telematic assemblage and this inhibits a taking for granted of its unified nature. Whereas in a pre-digitalized technological culture, the virtuality of risks still needed to be argued, usually against realism, for telematics the virtual has already taken over. As telematics fundamentally disrupts the categories of modern thought, risks also become inherently destabilized by the sheer force of virtualization.

A simple example would be the 'I Love You' computer virus which, apart from directly jamming network systems, also caused immense indirect problems as system operators and webspinners across the world had to generate warning messages and upgrade virus detection systems. This is an example of a risk that is able to reproduce itself not simply through failures of regulation and security, but exactly because of the risk-aversion ethos already put in place by the systems themselves. Once risks are able to work through successful risk-management systems and use them to proliferate themselves, that is, once they are not only able to act but also to anticipate and learn, we could speak of cyberrisks. The notion of virtual risk has been used to implode the realist/ constructionist distinction between the real and the representational. Beck's notion of risk involves a complex alignment between very particular and very exclusive modes of signification (most notably those of 'applied' and 'pure' science), with financial, economic, judicial, political and administrative institutional systems of representation. Beck's rather exclusive focus on 'institutional' risk perceptions must be supplemented with the popular and allegorically organized moral and practical *sense* that risks make in everyday life settings. Risks are realised in the experience and appreciation of a variety of particular discursive and figurative alignments.

This necessitates that risks must have an *appeal* to an audience to be made sense of as risk. Risks thus involve an interpellation: a calling upon potential receivers to respond (without ever, of course, assuming only one possible response). In an age where television is capable of instantaneous worldwide dissemination of information to an almost limitless mass of receivers, the appeal of risk is formed in a cultural matrix which exceeds that of technoscience and institutional politics: it has the capacity to 'directly' intervene in the social density of everyday life. Not all 'risks' need to be mediated by either expert knowledge or locally embedded moral consciousness to operate as risks. Although the essence of risk always entails both an understanding of probability and intensity of harm in relation to a specific decision-making capacity, neither probability, nor intensity, nor harm, nor decision-making can be exhausted by their semantic operators. This is because semantics exclusively concern *logos,* that is, the reasoned logic of signification. Supplementing this *logos* is the possible availability of *nomos*—the practice of distributing intensities across a field of forces and intensities that is not its own. *Nomos* is distinct from *logos* in that whereas the latter always searches from the principle of reduction, of a reproduction of origin, *nomos* is a flow of deviations and irreducibility.

Nomos is the logic of escape without disappearance or sublation. It cannot be controlled by the sign. This *nomos,* or what some poststructuralists have called 'decentred subjects', generates the non-isomorphic deterritorialized supplement of virtual risks—their transformation into a material force that is no longer exhausted by its signifying semiotics. In short, cyberrisks—virtual risks of the telematic society—are self-replicating actors that are not mediated by consciousness. Virtual risks are the particular expressions of experiencing risk perception. These expressions concern the discursively mutated 'sense' that risks and hazards make in the everyday lives of human beings, gods and technologies. Virtual risks are inscribed into the Abstract Machine of modern society where the signs in which they exist are being connected to 'a whole politics of the social field'. As a special case of virtual risks, cyberrisks are multiplicities from the start; they exist primarily as 'determinations' that cannot be reduced to any code existing beyond itself. The code itself is the pathological agent. There is no 'pure form' of cyberrisks; they are 'only' mutations and deviations. Cyberrisks have no 'history'. As they do not unfold alongside a narrative path in their revelation, they cannot be traced backwards to some mythical origin. Cyberrisks exist only in dissemination. Whereas virtual risks can still be traced (as objects of visualization), cyberrisks can be mapped only because its flows are multiple. The map is open and connectable in all of its dimensions; it is detachable, reversible, susceptible to constant modification. It can be torn, reversed, adapted to any kind of mounting, reworked by an individual, group, or social formation... A map has multiple entryways, as opposed to the tracing, which always comes back 'to the same'.

The map has to do with performance, whereas the tracing always involves an alleged 'competence'. The obsession in network computing with distinguishing between viruses and hoaxes can be seen as a denial of cyberrisks. Trying to trace a cyberrisk to its alleged origin in the mind of a criminal-obnoxious 'geek' may be a favoured pastime of criminal investigators and media pundits; it merely serves to mystify the nature of cyberrisk, which is never ascertained by origin but by its dissemination. Hoaxes are like viruses; they may also use email alongside television broadcasting to spread; they are as irritating and—because they proliferate autonomously from, and in fact exactly because of, the security concepts of 'safe computing'—they are as effective. The sheer amount of viruses, trojans, worms and hoaxes, combined with the speed at which they travel and the inability of virus-detection systems to stay ahead, makes 'tracing' a rather futile activity. The fact that cyberrisks can spread from and to any part of a world still carved up by national borders means that the 'policing' of cyberrisks can only work through deterrence.

The interplay between the fabrication of cyberrisks and the forensics of their tracing becomes itself a strong motivator of the proliferation of risk sensibilities. Cyberrisks thus need to be appreciated both in their singularity and multiplicity. The question 'what is cyberrisk?' is rather futile. Instead, we have to ask a different question: 'what does *a* cyberrisk do?' We could ask this question about cyberrisks only in terms of their singular perfomativity (never in general terms, nor in their particular manifestations), in particular that of their pragmatics of sense-making. Successful mapping of a cyberrisk must resist the temptation to reduce 'it' to a signifier or code. It can be sensed only in its distributions (effects). It should simultaneously be clear that this cannot be an either/or enterprise. Whenever we engage with cyberrisk reflexively, we are transcoding the matter-at-hand. Transcoding works both ways—it mediates between the virtual and the cybernetic. The reduction of 'cyberrisk' to a sign or code through signification, is thus simultaneously (in a supplementary sense) an embellishment. In the same movement, the signifying semiosis of 'virtual risks' sets into work a new information-processing *poiesis,* capable of generating an excessive induction of singularities into the symbolic order. Computer viruses are predisposed to escape the jurisdiction of their creators, dispersed by integrated circuitry, and capable of utilizing previously accumulated signifying material for their own explosive replication, and in these senses might be seen as the archetype of the superconductive event. The computer virus as the archetypical cyberrisk has been at the centre of attention of many media, not only news.

Both popular science and science fiction have embraced the immense apocalyptic sensibility of cyberrisks to sketch futures totally out of control because of self-replicating artificial intelligence. Books and films about the computer age, and especially those involving computer viruses are usually framed in a dystopic setting. Like the misanthropic apocalypses of emergent viruses as nature's revenge, they tend to sketch the future in very dark colours. However, unlike the nature's revenge domesday scenario, the location of cyberrisks is unequivocally the technological culture of the west. This culture is described as dehumanizing and alienating. What characterizes many of these accounts is a connection between cyberrisk and dystopia that is mediated by intentional political violence against humanity itself. That is, cyberrisks manifest themselves in popular cultural media as the epitome of conspiracy theses.

THE JOUISSANCE OF VIRAL TERRORISM

In *Virus*—one of the first monographs on computer viruses—Allan Lundell starts his exploration into the world of computer viruses with the case of a highly virulent internet virus that came to existence in 1988. It was designed to jam the entire system worldwide. It failed because of a programming 'error', which Lundell suggests may have been deliberately designed. What allowed this virus to be so powerful was its ability to break through the various security devices that were designed to protect the system from such nasty intrusions. It is the story of a rapidly spreading digital bug that was programmed to explore the gaps in electronic mailing systems. As a peculiar twist to this story, Lundell adds a genealogical note: the designer of this virus, Robert Morris Jr, is the son of a world famous expert in computer security, Robert Morris Sr. To programme such a virus is both seen as a criminal act and a highly skilful accomplishment—the aesthetics of the sublime are those of a radical evil. The passion that both father and son Morris have for computing is a libidinal economy that is simultaneously inscribed in the paranoid warfare economy of the state apparatus.

Lundell finds a closure in this story by pointing out that Morris Jr's act has led to a wealth of insights in the vulnerability of networked computing systems as well as in the operations of such skilfully designed viruses. The anxiety of risk of failure is thus offset by a will to know and a belief that complete security and regulation can be achieved. However, cyberrisks are not that easily put off by Oedipal desire as they operate in a rather different libidinal economy. Since Lundell's now very dated book, our world has become increasingly dependent on computing systems and telematics. In the mean time, a plethora of new viruses have appeared, most notably Michelangelo', 'Melissa', 'I Love You', 'Red Alert' and 'Nimda'. The most feared viruses are those that use email attachments, appropriating a device designed to make telematics more user-friendly. With the global domination of Microsoft over the design and standardization of software applications, telematic systems have become extremely vulnerable to cyberrisks because of the relentless drive to acceleration and increased compatibility. Indeed, nomadic forces of cyberterrorism find it easy to travel through the deterritorialized spaces of global capitalism. However, for every virus, there have been hundreds of hoaxes. Perhaps one of the most famous hoaxes has been the 'Millennium Bug' or Y2K. During the mid-1990s, noises emerged, primarily in the form of rumours, that computing systems could experience a breakdown at the turn of the millennium because of the way in which computers kept dates.

The year was marked by only two digits, to save memory space on the already overcrowded original memory boards, a practice that was simply sustained even when no longer necessary. At the year 2000, when those digits turned to 00, all hell would break loose, since computing systems would act as if it was 1900. Billions were spent on rewriting the configurations and replacing computing systems with ones that were Y2K-compatible. Then, when the clocks turned to 00.00 on 1 January 2000, nothing happened. Even computing systems that had not been upgraded, such as those controlling Russia's nuclear plants, functioned 'normally'. Instrumental fidelity did not fail us, and the world breathed a sigh of relief. The point however is not that it was a hoax, but that no one, not even computer experts, was able to either verify or deny the rumours. It also showed how badly prepared our institutions actually were, and how slow they reacted to such potentially lethal risks. Above all, it showed how deeply entrenched microchips and data-processing have become in our daily lives. People were invited to make connections not only between the hardware of their PCs, cars, video recorders, televisions and refrigerators, but also between their own private systems and those of a collective nature such as banks, traffic control, hospitals, public transport, aviation, nuclear plants, policing, etc. And this had to be thought of as on a global scale. Hence the situation at the Russian nuclear plants was as significant as whether we could still draw money from a cash machine. Indeed, people were encouraged to reflect on the global complexity of connectivity and to see how fragile they have become in the face of risks.

This is no hoax but a truly transformative event in contemporary technological culture. Now that the smoke has cleared, everything seems to retract rapidly into the former segregated conceptions of modern life. Apparently, no lesson has been learned. Technological culture has indeed been shown to be quite resilient. The adhocracy of the quick-fix still rules supreme. Yet, a sense of global interconnectedness has not entirely disappeared. Internet viruses occasionally strike at the heart of telematics, and point towards its inherent precariousness. Internet security has continued to be a top priority for governance, commerce and the military, and initiatives are being taken to set up a worldwide information regulation policing system; something currently carried out de facto by the Federal Bureau of Investigation (FBI) and—since 11 September 2001—by the US Army. Risk remains associated with anxiety and paranoia. In the world of telematics, cyberrisks are particularly uncanny in the way in which they seem to bring to life medieval sensibilities about the dark side of human being. Such ideas can be found in two relatively recent but not quite so successful films: *Cyberjack* starring Michael Dudihoff; and *Cybertech P.D.* starring Lorenzo Lamas.

In *Cyberjack,* a multinational computer company is involved in the development of a computer virus aimed to defend computing systems against all kinds of unwanted intrusions. The company is then taken over by a group of terrorists who seize the virus to create chaos in order to control the world. Michael Dudihoff, the hero, manages to kill them all and save the world. The film opens with the following quote that is allegedly Stephen Hawking's: A computer virus should be considered a form of life; that the only form we have created so far is purely destructive. We've created life in our own image.

Although, in the narrative, risk is predominantly framed in terms of 'bad' people seizing a telematic system through a malicious computer virus, cyberrisk constitutes the very possibility of this taking place. As a form of life, the computer virus had already enrolled so-called cyberterrorists to engender its autopoietic take-off. Similar to the conducive role played by viruses in genetic engineering, computer viruses are a central force in telematic systems, they are the conductors of transcoding, moving from one system to the next, rewriting the codes as they move along. It is thus no wonder that cyberrisks are a source of great concern, even panic. They signify the limits of our technological culture through the excessive nature of their pathogenicity. However, because they are often associated with and even reduced to specific wilful individuals, who are motivated by evil, cyberrisks are also recuperated by a morally coded technoscience of information forensics and the governmentality of information policing. Like Dudihoffs bold hero, these agents of modern technological culture may rescue us from our damnation. Because cyberrisks are often denied any agency apart from their creator's—that is because we fail to understand cyberrisks as cyberrisks—we promote a false sense of security and endorse a misplaced faith in securing and regulating systems.

Nowhere has this become more clear in the events running up to the attack on the World Trade Centre and its aftermath, in the anthrax scare of October 2001. Whereas in *Cyberjack* the threat comes from 'bad guys' using 'neutral' technology, in *Cybertech P.D.* the threat is technology itself. Virtual reality (VR) is placed on the same continuum as pornography and drug use. The key word here is 'addiction'—the institutionalization of mental and physical dependency upon external stimuli. *Cybertech P.D.* is thus based on a more traditional police narrative and involves the usual mixture of sex, drugs and violence. It features a cyber-policeman played by Lorenzo Lamas who is on a quest to arrest a former drug-cartel baron, now VR-entrepreneur. The emphasis here is not as much on viral cyberrisks but on a general sense of corruption and violation that virtual realities engender, in particular when bio- and cybertechnologies are interconnected. In *Cybertech P.D.* the hero-cop is technologically enhanced and connected—via a Gibsonesque sort of sim-stim (simulation stimulant)—connected to the Cybertech police force. However, his heroic status solely depends on his ability to operate outside the technosphere. He must remain an individual and resist the seductions of virtual sex and cyberdrugs. His moral code stems from his romanticized integral-corporeal embodiment (humanism). Hence the real threat comes from the cyber-bourgeoisie-mafia who have now seized upon biotechnology to clone actual human beings (mostly women of course) and turn them into sex objects. If this transgression of the virtual is accomplished, there are no safe havens for human corporeal integrity—all is corrupted.

In the climax, the hero injects himself with a superdrug to overcome the impossible, defeats the bad guys but seems to be lost forever, as he himself is now corrupted beyond repair. However, through embodied desire for natural romantic love, he is 'rescued' by his female counterpart and able to resist technological corruption, thereby re-establishing Human Nature over (demonic) Techne. Addictions are cyberrisks because they work as a virus. They engender a parasitical dependency relationship in which agency is surrendered to an externalized force. The integral body is corrupted and foreign agents take over. In medieval times, such phenomena were often associated with demons; angels of the Evil One, who would enter a body, often through sin, and take control. The term demons is derived from the Greek *daimones*. St Justin, an early Christian thinker, took this Greek term, which at that time still had a neutral meaning referring to spiritual energy and associated it exclusively with the powers of Satan. It is not surprising that addictive behaviour was often seen as demonic. Addicts act as if possessed by an external force, to a degree it becomes self-destructive. The loss of personal integrity is also coupled with the loss of free will and the ability to make correct judgements. Addictions prominently features among cyberrisks of the 1990s.

Varying from addictions to games, chat rooms or internet pornography, ICTs are often demonized in the sphere of leisure and consumption. Hence, a recent outcry over paedophiles using internet chat rooms to contact underage teenagers is merely an example of the way in which virtual reality is conceived as an inherently dangerous and easily corrupted space, situated beyond morality and security. This anxiety goes beyond the fear of paedophiles obtaining virtual presence in the safe haven of a child's bedroom; it extends into the fear of the child her- or himself being corrupted by it. Both *Cyberjack* and *Cybertech* P.D. display a deeply modernist neurosis as their founding psycho-semiotic constituency. Both express an anxiety over the vulnerability of the human condition in the world of cybernetics. They thereby echo a whole genre of dystopian films about cyborgs and cybernetics, such as *Terminator, Robocop* (and their sequels), *Strange Days, Bladerunner,* etc., which all evolve around an association between telematics and pathologized human evil, in particular megalomania. Holland is right to point out that a fascination with cyborg films is often induced by paranoia, but Nigel Clark's assertion (derived from Baudrillard) that we are actually fascinated by the creation of a catastrophe should equally remind us that not all apocalypses are dystopic, and not all dystopias are anticipated by paranoia. Obviously, such films, like Lundell's book, are not outside the genre and ethos of the socio-cultural climate in which they were engendered.

They should thus not be seen as anything but representations of our own relationship to the present, mediated via a construction of 'futurity'. That viral terrorism and biotechnological corruption threaten the integrity of the system-as-we-know-it is left unproblematic; it points to a fundamental psychopathological disorder of modernity itself: neurosis. Lundell's *virus* is stifled with this modernist paranoid neurosis. However, his accounts of the various viruses, worms and trojans that plagued the computer world in the 1980s also include a fascination with destructive evil and the ability of peculiar cybercowboys to trace and neutralize them. Moreover, his book offers itself as a *pharmakon,* holding both the poison and the medicine. Indeed, like a television evangelist, Lundell paints both the apocalypse and the possibility of redemption ('buy this book'). The fascination with the 'preprogrammed catastrophe' constitutes the enigma of his encounter and becomes further apparent through the numerous hints at the genius of particular virusprogrammers. For Lundell, viral design involves a kind of narcissistic aesthetics—computer nerds create viruses to show how clever they are by testing their masculine egos in a new and powerful environment.

The quality of the virus is interpreted in terms of the effects it has on users, in particular its velocity and degree of infectiousness, or 'epidemic of simulation'. Thus, Lundell leaves us with some rays of hope—there are remedies against computer viruses. During the late 1980s and early 1990s, many network operators tried to cultivate an ethos of 'safe computing'—often in close analogy to the AIDS-hype which by then had reached its peak. Not buying illegally copied software or copying software from message boards, never using disks or opening email attachments without checking them for viruses, installing virus scanners on mainframes and PCs, all were part of a newly institutionalized anti-virus campaign. Yet, despite all these efforts, many systems proved highly vulnerable to viral epidemics. The computer network of the University of Wales in Cardiff experienced three main outbreaks of the 'Trajectory' virus between March 1995 and March 1996. By then, viral scanners were already widely used and installed to automatically monitor all the servers on the network. Apparently these measures were insufficient as the virus was too quick and too virulent, too subtle and too effective, for such institutionalized safetynets to function properly.

Of course one may attribute such pairings of apocalypse and redemption to a genre of modern fiction-writing; but, as has become clear since 1990, it is the scientists who are now the main contributors to the panic scenes of risk society. Nuclear technology, genetics, waste management, AIDS, antibiotics and computer viruses are all incorporated into stories written in anxiety and fear on the one hand, and an optimistic faith in an ultimate victory of reason on the other hand. As the apocalypsis starts to gain momentum, reason becomes increasingly revolutionary, that is, the 'solutions' become increasingly 'unreasonable'. To reduce these apocalypse/redemption pairings exclusively to forms of virtual risks, that is to forms of popular cultural expressions that have no index of effectivity outside 'the' mediascape, is to miss a vital point: they have an innate sense of en-presenting themselves beyond the technologies of visualization and signification upon which virtual risks rely. It is the connectivity between the desiring machines of the apocalypse and those of the unreasonable that engender the cyberrisks we are subjected to today.

NEUTRALIZE THE NEUROSIS!

What is so typical of the risk society as described by Beck is that it suffers from a deeply entrenched and very modernist form of neurosis. The fate of modernization is that of unforeseen and unforeseeable consequences that turn against the movement itself. Risk society marks that moment when such a fate is recognized as imminent to the systems that are designed to control and overcome it (mistaking fate for destiny). Why, then, do so many (social) scientists discuss the ecological fate of risk society in terms of institutionalizing risk management? Why, then, is so much energy invested into developing a new rationality to counter the problems of modernization, in terms of 'sustainable development'? Sustainability may be something that has less to do with management and control, but instead with letting go, with breaking down the artefacts and barriers within which the state apparatus has entrenched itself against attacks from the margins. Perhaps the most vital lesson can be learned from the viruses themselves.

They tell the story of the eternal return. In a world in which the survival of the human species is no longer a certainty—due to a seriously disrupted ecosystem as a result of which they are no longer capable of providing enough oxygen, water, carbohydrates, protein, vitamins and minerals without lethal intoxications or infections with encephalopathic prions and such—we may not want to wait for institutionalized reflexivity to pull us out of the sewer in which we find ourselves stuck while busy creating it. The media-science connection has made this all too clear—we are doomed, at least for the time being. Politics, law, governance and commerce are not going to rescue us, as their timescales are far too short to deal with the organization of the risks and hazards they effectuate. Because the apocalypse looms large, we have to turn to the revolutionary desire of the unreasonable, the aesthetics of the nomadic war machine. This is the lesson of the cyborg that Donna Haraway points us towards in her manifesto.

7

Scarcity of the Radio Frequency Spectrum

The electromagnetic spectrum describes the oscillations of electromagnetic waves that propagate at, or near, the speed of light through space or conductors. Frequency is the number of oscillations, or waves, per second and is measured in Hertz (Hz). Wavelength is the distance between crests of an electromagnetic wave, measured in meters. All electromagnetic waves—radio, television, radar, infrared, visible light, laser, ultraviolet, X-rays, and gamma rays—are functionally identical and travel through a perfect vacuum at precisely the speed of light—300,000 kilometers/second (km/sec) or 187,500 miles/second (rounded to 186,000 miles/second). The types of electromagnetic waves are distinguished only by frequency, which is proportional to wavelength: higher frequencies have shorter wavelengths. Over the entire electromagnetic spectrum, both frequencies and wavelengths vary over twenty-eight orders of magnitude from infrasonic waves to cosmic rays.

RADIO FREQUENCIES

The radio frequency (RF) spectrum is ubiquitous in nature and physically accessible to anyone with rudimentary technology. Its universal accessibility and nonexclusivity, paradoxically, invite interference, which is the primary limitation and constraint on its use for telecommunications.

Too many communication transmissions at a single frequency interfere with each other, by superimposing one on another, or by overpowering the weaker signals. This interference places physical limits on the capacity of any band of RF frequencies to carry information. To minimize interference, the International Telecommunications Union (ITU) has designated specific bands of frequencies over the entire RF spectrum for particular applications and users. Current telecommunication users of the RF spectrum include fixed and mobile satellites, terrestrial radio relay (microwave), terrestrial radio and television broadcast, air-to-ground communications, wireless cable and fixed television, amateur radio, emergency mobile radio, cellular radiotelephone, paging services, wireless private business exchanges (PBX), local area network (LAN) services, disaster warning networks, military communications, extraterrestrial communications, and other market-driven applications.

PROPERTIES OF THE RF SPECTRUM

The variability and uneven efficiency for various types of telecommunication services further constrains the RF spectrum. The utility and value of the electromagnetic spectrum for relaying information are determined by the physical properties of the different spectrum bands, the propagation patterns of RF waves, and the technologies involved. At low frequencies, the long wavelengths allow radio waves to bend around obstacles—broadcast need not be line-of-sight—and to penetrate atmospheric aerosols and liquids without distortion. Thus, at low frequencies, antenna systems need not be very precise in direction and require little power to send transmissions. For these and other reasons, lower frequencies are technically and economically preferable to higher frequencies. Higher frequencies, however, have a powerful advantage. The capacity of RF waves to carry information increases with frequency, since more waves per unit of time provide more sites to modulate. Some broadband signals simply will not fit into the bandwidths of low frequencies. "Broad band services imply the use of the upper frequencies and the need to access very broad segments of the spectrum." Faster oscillation at higher frequencies requires more energy, which must be provided by some power source. The higher energy levels of higher frequencies make the RF waves more difficult to control in a conductor or a transistor, or even radiating through space. The difficulty and cost of modulating, amplifying, transmitting, and receiving RF waves consequently increases as frequency rises.

As demand for the simple, cheap, lower frequencies to carry first-generation geosynchronous earth orbit (GEO) satellite communications increased, technological innovations allowed increased use of higher-frequency bands with greater information-relaying capabilities, but at correspondingly higher cost. Users of lower-frequency RF carriers still prefer to avoid the increased costs of higher frequencies, unless the technical, economic, or operational advantages are clear. Another powerful dissuasion to shift to higher frequencies is the extensive investment in C-band and Ku-band infrastructure. Radio waves propagate differently through various environments and show differing efficiencies at different frequencies. Although they carry less information than higher frequencies, low frequencies penetrate even dense environments and flow around many small obstacles. Because of their shorter wavelengths, transmissions at higher frequencies experience more unwanted absorption by precipitation and airborne aerosols, which progressively degrades communications performance. Frequencies above 4 megahertz (MHz) and below 27 MHz are blocked or reflected by the earth's ionosphere and are, thus, not preferred for satellite use. Signals at frequencies above about 10 gigahertz (GHz) are seriously degraded by heavy rainfall between transmitter and receiver. Waves at very low frequencies—below about 10 kilohertz (kHz)—are reflected by ionized gases, atmospheric disturbances, electromagnetic waves, and cosmic rays, as well as other radio signals and are not useful for atmospheric communication. The atmosphere is broadly transparent to waves at frequencies between about 30 MHz and about 300 GHz.

Frequencies above about 300 GHz are scattered and attenuated by atmospheric aerosols, electromagnetic waves, cosmic rays, and other radiation, as well as by electromagnetic interference. Light waves—including laser beams—with even higher frequencies, are absorbed by dust clouds and aerosols and are scattered by atmospheric turbulence. In the void of space, with no atmospheric interference, it is possible to relay signals between satellites by modulated laser beams, with the optical spectrum's vastly greater transmission capacity and speed. The Japanese Space Agency, the American Jet Propulsion Laboratory, and several commercial firms are developing these concepts of wireless laser communications.

USING THE RF SPECTRUM

Wireless communicators can use the RF spectrum as a channel for communication by intentionally deforming, or modulating, electromagneticwaves with either an analog signal or a digital bit stream. A transmitting RF antenna radiates the modulated waves into free space, without a wire or other conductor, where they are collected by receiving antennae. The receiving equipment demodulates, or interprets, the deformations in the waves and thereby extracts the information, message, data, or bit stream transmitted by the sender. Analog signals are then delivered to the listening audience, or bit streams are demultiplexed and decompressed into the original message. Since RF energy spreads very rapidly into a large beam cone, or footprint, the clear advantage of this physical process for wireless RF telecommunication is that anyone can receive the signal in the broadcast area whether they are at a fixed location or in a mobile unit. This allows new links to be established at a moment's notice since physical links do not have to be installed. The disadvantage is that the "frequency" is used up for the entire transmitting zone in contrast to a wire transmission that restricts the signal to inside the conduit. It also means limits on privacy of communications, on priority of access, and on exclusivity of use. Table summarizes the RF frequencies and bands normally used for wireless telecommunications, with some common applications at the frequency bands preferred for telecommunication applications.

For radio telecommunications, the designated frequency ranges are from 10,000 Hz (10 kilohertz, or 10 kHz) to 300,000,000,000 Hz (300 gigahertz, or 300 GHz). Above the UHF bands (300-3000 megahertz, or MHz) the microwave bands extend to 30 GHz and the millimeter bands to 300 GHz. Like all electromagnetic waves, radio waves can also be measured as the distance between crests of successive waves; these distances in the RF spectrum range from more than 30,000 meters to less than 1 millimeter. The relation between frequency and wavelength is inverse; their product is the speed of light. In some of the frequency bands in the RF spectrum in some places on earth, humanity's transmissions of information are approaching maximum capacity.

Much of the wireless electromagnetic infrastructure—modulation techniques, amplification processes, transmission isolation, high data rates, antenna systems, retransmission and relay methods, message encoding and protocols—has been rebuilt many times already and will be rebuilt again as new, innovative technologies and clever transmission techniques appear. But additional spectrum is not available for any significant expansion of transmission capacity. Many telecommunication applications simply do not work outside a narrow band of frequencies. Voice broadcast is most effective at frequencies between about 1 GHz and about 30 GHz. Below that range, antenna systems become large, expensive, and unwieldy; above it, signals are easily blocked and attenuated by obstacles, buildings, foliage, and even rain or fog. The physics of electromagnetic propagation requires that cellular telephony—like most voice communications—occur in this spectral range to be effective, efficient, or even useful. For most users of low voice broadcast frequencies—VHF, UHF, C-band, and Ku-band—the problem is obvious: the increase of data and video traffic—the so-called broadband applications—at the same frequencies. If most data and video traffic could be removed from C- and Ku-bands—by using higher frequency bands—the remaining voice traffic would flow more smoothly without congestion, with less interference, and with better quality. Moving to higher frequencies—with higher transmission capacities and speeds—for broadcasting data and video subjects these signals, however, to atmospheric attenuation and possible delay.

Traffic that is not time sensitive, interactive, or duplex could readily migrate to higher frequencies, although clearly much traffic must—for economic, physical, or even political reasons-remain at voice broadcast frequencies. Since the technologies needed for migration and for reconfiguration of satellites, antenna systems, transmitters, and receivers are neither expensive nor innovative, the major obstacle to migration seems to be either traditional habit or the pressure of market forces. The comfortable conveniences of transmitting all broadcast traffic as if it were duplex voice and of broadcasting data and video intended for a single recipient, rather than using point-to-point channels, have created congestion in voice broadcast bands and are beginning to increase the cost of transmission capacity in those bands. It may be these same market forces that ultimately force migration of some traffic to higher frequencies. Despite its ubiquity as a global resource, some parts of the spectrum – currently those preferred for cellular telephony and mobile services, which experience low weather attenuation and high penetration: VHF, UHF, C-band, and Ku-band—can also be measured in dollars.

All wireless communication services, with the *exceptions of free-space communications at infrared and optical frequencies*, require national and international allocations of bands of the RF spectrum. Another notable U.S. exception is the so-called "part 15" band available in the United States without allocation or license for use at such low power that propagation is confined within a single building. These "part 15" applications include wireless switchboards, infrared links between computers and modems, remote controls for television or video devices, and intrusion detectors. The recent phenomenon of frequency auctions reflects the scarcity of RF spectrum and emphasizes the constraints imposed on telecommunications growth by both the physical characteristics of the RF spectrum and the politicaleconomic-diplomatic efforts to allocate frequency bands either to uses or to users. In response to the Omnibus Budget Reconciliation Act of 1993, the U.S. Federal Communications Commission (FCC) and several other national regulatory authorities have begun regularly to auction bits of the RF spectrum for particular applications – notably personal communications service (PCS), commercial mobile radio services (CMRS) and private mobile radio services (PMRS). The price of such an asset can reach $100 million for a single megahertz of frequency. National authority to conduct such auctions devolves from the prerogative of the ITU to allocate bits of the spectrum every five years to national governments to use to meet their own requirements.

COMMUNICATION SATELLITES

For several generations, communication satellites have provided a prominent channel for relaying and distributing information – messages – in the forms of voice and data telephony, and television or video signals around the world. Although satellite networks operate in many roles and functions, the ITU recognizes seventeen separate services that satellites – in any orbit – provide. These ITU services vary greatly in commercial value and degree of use.

- Fixed satellite services (FSS)
- Broadcast satellite services (BSS), commonly known as direct broadcast services (DBS)
- Broadcast satellite service for radio (BSSR)
- Radio determination satellite service (RDSS)
- Radio navigation satellite service (RNSS)
- Aeronautical radio navigation satellite
- Services (ARNSS) Maritime radio navigation satellite service (MRNSS)
- Maritime mobile satellite service (MMSS)
- Intersatellite service (ISS), also known as intersatellite links (ISL)

- Mobile satellite services (MSS)
- Aeronautical mobile satellite services (AMSS)
- Land mobile satellite services (LMSS)
- Amateur satellite service (ASS)
- Meteorological satellite service (MSS)
- Space operation service (SOS)
- Space research service (SRS)

With some satellites operating in several networks, the ITU listed, in September 1997, 819 GEO satellite networks as "in use." Proposed new networks in GEO, MEO, and LEO "notified" to the ITU Radiocommunication Bureau through 1997 promise a further 439.

FIRST GENERATION

First-generation communication satellite operators began to place large satellites in the GEO to provide satellite services at radio frequencies in the early 1960s. Although satellite operators may provide services in many regions of the RF spectrum, commercial satellite communications through the GEO are constrained – by a combination of physics and regulation – to three bands of frequencies: the C-band (6 GHz uplink to satellite and 4 GHz downlink to ground stations); Ku-band (14 GHz uplink and 12 GHz downlink); and Ka-band (30 GHz uplink and 20 GHz downlink). The C-band is also heavily used by terrestrial communicators and radar operators. Because of its indifference to weather and the extensive experience and infrastructure with it, C-band is the most intensely used of the RF frequency spectrum bands. In addition to point-to-point and broadcast services, GEO satellites provide circuits between communicators operating mobile ground stations on terrestrial, airborne, or sea vehicles. This form of FSS uses frequencies in the L-band (1-2 GHz) and S-band (2-4 GHz). Mobile military users of FSS use designated frequencies (235-399 MHz) in the upper VHF (30-300 MHz) and UHF (300-1,000 MHz) ranges. Stationary military users operate in the X-band (8-12 GHz) and purchase commercial capacity in C-, Ku-, and Ka-bands.

SECOND GENERATION

Second-generation communication satellites will occupy lower orbits and convert conventional analog communications signals to streams of digital bits before transmission. Operating in LEO or MEO, constellations of these modern, light, small second-generation RF communication satellites will provide MSS at radio frequencies that are higher (Ka-band) and lower (L-band) than those associated with first-generation FSS (C-band and Ku-band). Most secondgeneration MEO and GEO telecommunication satellites have selected operating frequencies in L-band (1.5-2.7 GHz), with a concentration in the range of 1.50-1.62 GHz, with some uplinks at 2.4-2.5 GHz. The many LEO voicemessaging and paging systems prefer slightly higher uplink frequencies (1.6 GHz) with downlinks generally at 2.4 GHz-2.5 GHz. Salient exceptions are the Iridium intersatellite links at 22.35-23.55 GHz (Ka-band) and Teledesic operating uplinks and downlinks in Ka-band. The so-called telegram, or store-and-forward, systems—OrbComm, StarSys, and LeoStar—all operate in VHF and P-band. In addition to extending spectrum usage to higher and lower frequencies, the new second-generation MSS systems incorporate innovative radio engineering and antenna technologies to increase the efficiency of their use of the RF spectrum. Although the early second-generation systems are focusing on current high-demand markets for duplex voice telephony, telefacsimile, and data transmission, many operators also propose to provide additional—or eventual-capacity for video traffic. Some—notably, Teledesic—even propose to provide basic wireless telephone service. To provide affordable high-capacity, high-speed services—interactive multimedia, Internet connection, and television—at speeds tolerable to modern computers through small desktop or handheld antenna systems, second-generation satellite systems or first-generation GEO systems must operate at the highest frequencies possible with very high power. Even with intensive frequency reuse, phased array antenna systems, polarization, and other sophisticated techniques, second-generation systems consume huge slices of RF spectrum. Providing high-capacity services at low frequencies will inevitably congest those frequencies and force some applications to migrate either to fibreoptic networks or to higher frequencies.

RADIO TECHNOLOGY

Although congestion and interference are the primary physical constraints on use of the RF spectrum, the economics of technical progress also limit access. Technological innovation in design or function of ground stations, receivers, and transmitters, which generally leads to lower cost, motivates more users to transmit more information using RF carrier waves through first-generation GEO satellites. The increased demand increases congestion, since radio technology generally does not create new capacity, but rather increases the efficiency of using existing capacity. The result historically is more frequency congestion, despite greater efficiency in using it. This eventually requires yet another more sophisticated, and more expensive, technological innovation to deal with it.

EXPANSION INTO HIGHER FREQUENCIES

Because high frequencies can carry more data than low frequencies, those used for telecommunication have increased from tens of MHz at midcentury to nearly 100 GHz at century's end. Technology has expanded the exploitable RF spectrum for satellite communications by using higher frequencies—Ku- and Kabands—with greater information-relaying capabilities. Experiments aboard the Japanese Engineering Test Satellite (JETS) VI have shown that atmospheric and space communication at laser frequencies are definitely feasible. Laser telecommunication systems are especially well suited for carrying very heavy streams of traffic that require huge amounts of transmission capacity. At higher frequencies, the shorter wavelengths below about a millimeter —infrared and optical—are easily absorbed by the atmosphere. Sensitivity of transmissions to weather interference—precipitation attenuation, atmospheric heat scintillations, and random ionization distortions—in these higher frequency bands has spurred development of higher power RF transmitters that can often compensate for rainfall losses in up- and downlinks between earth and satellites. Because of their vulnerability to unpredictable interruption, however, these higher frequencies offer little utility for many interactive, multimedia applications.

ISOLATION AND REUSE

A second approach has stressed more efficient use of the GEO and reuse of C-band and Ku-band frequencies by isolating receivers from interference. The several techniques for electromagnetic isolation include: orbital isolation, spatial isolation, geographic isolation, polarization isolation, and temporal isolation. Increasing the space between satellites allows a ground station to see only one signal source in its range of vision, or beam width. Orbital isolation, however, decreases the number of satellites that the GEO can accommodate and increases the difficulties of tracking moving satellites in lower orbits. Spatial electromagnetic isolation is the result of concentrating transmission broadcast cones into narrow spot beams that reduce a satellite's footprint.

Using very high power to boost the effective isotropic radiated power (EIRP), large modern GEO satellites can carry up to 120,000 voice circuits in a near point-to point mode. Creating many small, isolated zones, or spot beams, for wireless use of the same RF frequency allows reuse of RF frequencies and effectively increases the speed and effective capacity of a satellite broadcast system. Geographic isolation of ground stations—through distance between them-permits reuse of frequencies by satellites located in the same section of the orbital arc. Sophisticated station-keeping techniques allow collocating satellites and intense use of preferred portions of the arc. Isolation of ground stations loses its utility in systems with mobile ground receivers not controlled by a system operator (Iridium's global cellular telephony). Personal communication and digital cellular telephone systems divide the allocated satellite frequencies into subsets. A single subset serves each cell site, which must be geographically isolated from other cell sites that are using the same subset frequency. Another approach to isolation is use of the new phased array antenna system. Multiple transmitting or receiving elements arranged in a stable geometric array either send a grid of electronically formed discrete radio beams or receive signals from only specific directions.

Phased array antenna systems can steer RF beams with accuracies of ½° and achieve spatial isolation through their directional selectivity by imposing delays or advances on signals moving from—or to-different elements of the array. Phased over both time and space by the array, signals are effectively isolated and do not interfere with each other. "Intensive reuse concepts are critical to the future growth and expansion of wireless [RF] service in the twenty-first century.... In the field of [RF] satellite communications new and innovative reuse techniques may prove critical to the future ability to deploy large numbers of geosynchronous, medium earth orbit and low earth orbit [RF] satellites needed to respond to new market demands."

POLARIZATION DISCRIMINATION

Sensitive adjustments and complex coordination of transmitting and receiving antenna elements can generate polarization isolation of RF signals. Radio waves with vertical or horizontal polarity and with left-hand or right-hand polarity may be produced, transmitted, and received simultaneously and independently on a single frequency. Polarization isolation, which is most effective at C-band, increases the sensitivity of Ku-band and Ka-band to atmospheric interference.

SPREAD SPECTRUM

Usually used to supplement some error correction code, spread-spectrum technology exploits the direct variances of multipath fading and noise with frequency. Spread-spectrum techniques deliberately disperse the transmitted signal over a range of radio frequencies within a particular band: in effect this technique consumes all of the available capacity to guarantee reception of the signal. Time-phased spread-spectrum techniques switch the RF transmitter to a new frequency every few milliseconds and coordinate parallel switches in the receiver.

For data communications, sophisticated signal processing after reception allows reassembly of the signal. Direct sequenced spread spectrum techniques convert each separate data bit into several transmitted bits (chirps). The randomized chirp transmission sequence is known to both transmitter and receiver. Because this random sequencing operation generates large multiples of transmitted chirps for each bit of the original message, the actual transmission spreads over a wide range of frequencies—and consumes all of the available capacity—while minimizing multipath fading and external interference. Sophisticated signal processing at the receiver can reconstruct the original data with high fidelity.

OTHER EMERGING TECHNIQUES

Modulation, encoding, and multiplexing can provide connectivity among multiple ground stations. Digital time-division multiple access (TDMA, preferred by AT and T) or code-division multiple access (CDMA, preferred by Qualcomm, US West, and Bell Atlantic) techniques, rather than analog FDMA (frequencydivision multiple access) processes reduce intermodulation interference. By coordinated, ordered, sequential timing of ground stations transmissions and receptions, each message can occupy an entire transponder. Like other technologically innovative approaches, use of TDMA provides greater capacity, but only with greater complexity and cost, as well as increased requirements for maintenance and operating technicians. Onboard digital signal processing, signal regeneration, and space switching systems are further possibilities for increasing satellite RF transmission capacity. Digital data compression based on new standards that allow transmission of compressed data and voice signals substantially increases the capacity of a transponder.

THE RF SPECTRUM: EXPANDING OR CONGESTED?

"The clear pattern is that satellites for many different applications are gradually 'saturating' the available orbits as the demand for space services continues to grow even with the great popularity of fibre optic cable systems. This process of 'saturation' does not mean filling all of the physical space within an orbit, but rather, a saturation of available frequency bands." "The problem is congestion, regulation, and tariffs on all other competing bandwidths and technologies. This is such an overwhelming problem that even optimistic net-gurus are predicting major failures due to congestion in 5-10 years." Any assessment of the capacity of the combined resource of RF spectrum and geosynchronous orbit, with the enhancements of second-generation technology and lower orbits, to provide telecommunications must consider a broad panoply of technologies, physical features, continuously changing demands, and sensitive political interactions. In various contexts, the term *capacity* refers to the number of satellites in orbit; the amount of spectrum used; and the capabilities of ground stations to provide communication pathways. Few analysts have suggested that the size of the GEO fleet is approaching any physical limit. Even so, those GEO arcs relevant to high intensities and densities of traffic show electromagnetic congestion and competition for slots. Neither is the LEO-MEO-HEO fleet approaching any numerical limit.

Most analysts, however, agree on the constraints imposed by limitations on frequencies. Today the C-band is essentially filled—both physically and electromagnetically in GEO. The Ku-band is only slightly less congested. Only costly techniques for frequency reuse, digital compression, or polarization can expand satellite services in these bands. Frequency allocations for LEO systems around 2 GHz provide only low capacities that are subject to intense competition, as are allocations in other bands. The S-band (2.4-2.5 GHz) for satellite community antenna educational distribution is not fully utilized, and some of the military allocations in X-band (7-8 GHz) are not congested. A few DBS feeder and distribution links around 18 GHz are still available. For most FSS, MSS, and DBS applications, however, the preferred band for expansion is Ka-band (20-30 GHz). At higher frequencies, several experimental projects—the U.S. National Aeronautical and Space Administration (NASA)'s advanced communications technology satellite (ACTS), Italy's ITALSAT, Japan's JETS VI—are investigating techniques for overcoming rain attenuation at millimeter wave frequencies (Q- and V-bands). The commercialization of these still inconclusive experimental results as Teledesic's Mega-LEO, Norris's NORSTAR, Lockheed Martin's Astrolink, Loral's CyberStar, Motorola's Celestri, Morning Star, and Hughes's SpaceWay—all operating in Ka-band—reflects the industry's growing concern about scarce spectrum. "Currently proposed systems from the United States, Russia, and Europe, plus the anticipated filings for new international systems such as for the International Maritime Satellite Organization (INMARSAT) and Japan will more than claim all of the available spectrum even with intensive frequency reuse techniques."

In situations that require, or allow, one user's use of the RF spectrum to preempt that of another, the preempted users usually complain of—or at least note—insufficient capacity of the RF spectrum to carry their messages without interference. Although the physical bandwidth of particular radio frequencies may not change, the demand for noninterfering simultaneous broadcast transmissions exceeds the physical capacity of a particular frequency to carry them all. Users thus assess the "opportunity costs" of accepting preemption. For most commercial users—ultimately originators of messages—this opportunity cost is assessed in financial terms and compared with the market price of communication service. As constraints on capacity of particular frequencies have created relative economic values for various types of traffic, perceptions of scarce capacity have arisen not necessarily from technical or spatial considerations, but from electromagnetic and economic factors.

Many analysts assert that "whether through compression techniques or the deployment of fibre-optic technology, it is clear that the bandwidth used for transmission is becoming more and more plentiful." "Using digital technology and pushing it out effectively creates bandwidth." For the user willing and able to pay the price, technology can continue to provide adequate RF transmission capacity and satellite communication capacity, as the market preempts lower-value traffic. Presented with unacceptable opportunity costs, users unable or unwilling to pay either the price of preempting other users' traffic or the costs of implementing costly technological innovations will continue to suffer from insufficient capacity at their preferred frequencies. The impetus to migrate simplex data and video traffic to higher frequencies has heretofore focused on television. The minimal NTSC (National Television System Committee) television channel requires 6 MHz of spectra to carry audio and video—about 20 megabits per second, or 20 Mbps—which will not fit in the low frequency (LF) or midlevel frequency (MF) ranges, and only barely in the upper HF ranges. Despite the grossly inefficient use of RF spectra, early frequency allocations for NTSC television broadcast were in the VHF and UHF ranges (30 MHz-3 GHz).

The not-surprising result was increasing interference among the different NTSC signals. As the number of television broadcasters increased, relatively mild problems of interference—solved by various technological techniques—became serious congestion of these frequency bands. The TV solution—provided by national regulators—was to leave some channels unused in each metropolitan area and to allow the relatively minor requirements of low-power mobile radio and voice telephony to operate in these unassigned VHF-UHF channels. Although sanctified by decades of practice and vast infrastructures, today these allocations seem a wasteful, uneconomic use of these valuable spectra. The impending shift from NTSC to HDTV will place significantly greater demands on RF spectra used for broadcast television—both satellite and terrestrial. During the period of transition, both services will be broadcast simultaneously. The U.S. Federal

Communications Commission (FCC) decided in 1990 to provide to each of the about 1,500 U.S. broadcast television stations an additional channel for HDTV—the currently unused VHF-UHF channels that will not be needed for spectrum-efficient HDTV.

When the audiences for broadcast television had converted to HDTV—purchased new receivers—each station would discontinue NTSC and use the channel for additional HDTV channels. In May 1997, the FCC exchanged second channels for pledges to broadcast some HDTV programming when HDTV broadcast began in November 1998. "Most of the new channels lie within the UHF band; existing channels are to be given back once the [HDTV] transition is complete." The impetus to shift television—and other applications—to higher spectra reflects the emergence of fibre-optic cables that can provide television to consumers without any use of broadcast spectra at all. In addition, the higher frequencies of the millimeter and microwave bands (3-300 GHz) can carry far more broadcast television and other video and data applications on their greater bandwidths. The first higher band to be explored is Ka-band, which has an obvious advantage over S-, L-, P-, C- and Ku-bands for telecommunications: it carries more information. "Ka's higher frequency means that any chunk of data can be sent in less time, meaning faster transmissions." Perhaps just as significant for commercial operators, "[i]t is not as crowded as the other frequencies, and it enables smaller antennas." At these higher frequencies, however, broadcast television faces the huge costs of overcoming atmospheric attenuation, increased power requirements, and reducing satellite broadcast cones, as well as market resistance to replacing all existing commercial television infrastructure.

The U.S. federal allocation of frequencies to mobile communication services and the emerging concept of wireless personal communication services (PCS) reflect the effects of market demand on frequency valuation and allocation. They are also encouraging a general shift to digital services and data compression, which provide an important increase in RF transmission capacity at the low frequencies preferred for cellular and mobile telecommunications. The allocation of frequencies to accommodate this shift from wire applications to wireless services is being accelerated by new economic values associated with bandwidth, or capacity, through the auction process. Many analysts of first-generation telecommunications—led by Nicholas Negroponte of the Massachusetts Institute of Technology—predicted a dramatic shift of television—and any other broadband applications—to more efficient fibre-optic cable: the "Negroponte flip." Older, lower-value services (television) would—indeed, should—flip from expensive, limited RF broadcast frequencies to less costly cable transmission channels. In parallel, as higher-value transmissions (duplex telephony, paging) shifted to RF satellite carriers, Negroponte predicted the eventual disappearance of older inefficient, expensive media —copper wire, microwave radio, VHF-UHF fixed communications.

A more likely evolution is a less dramatic, rich but confused digital mixture of the old and the new. The "Pelton merge" suggests that each medium will continue with optical fibre, copper wire, terrestrial radio, and satellites carrying everything from voice to data to broadband multimedia and video. Ultimate users would demand access to video, data, and voice services from sophisticated mobile and handheld terminals. Each medium would carry virtually every signal for some portion of its route between origin and destination. Capacity, data rate, and quality would be those of the least-capable segment of the communication path—usually either copper wire or the RF spectrum. The technical heart of effective telecommunications would be the protocols for seamless connection between the various media with their immutable physical limitations, characteristics, and constraints. Although services and delivery media will shift frequency usage within the RF spectrum, the economic importance of installed infrastructure and plant, the continuing importance of deployed RF and satellite technology, and political inertia in reallocating RF bands can prevent any abrupt shifts. The evolution of the Pelton merge will require new data-conversion protocols and open systems standards that will allow data to flow seamlessly from copper wire to fibre-optic cable to RF transmissions to modulated laser beams in space.Whether they acknowledge the Negroponte flip or prefer the Pelton merge, many informed analysts and executives in both the wireless community and the telecommunication industry share a consensus that the RF spectrum is congested and the bandwidth is increasingly scarce. This generally accepted impression is developing into several clear trends.

- Technical Innovation: Development of increasingly costly technologies and engineering techniques to increase the capacities of the preferred C- and Kuband, especially through frequency reuse, modulation, encoding, multiplexing, complex patterns of frequency sharing, and digital compression.
- Reforms in the use of the RF spectrum: Exploding consumer demand and new technological innovations, inefficient historical patterns of RF spectrum usage, and obsolescence of several RF technologies that are forcing national and international managers of the RF spectrum into difficult reforms in the procedures, practices, and processes of frequency allocation.
- Economic optimization: The increasing-value of RF bandwidth as an economic resource; preemption or replacement of lower-value users by higher-value users and operators; determination of RF spectra allocation and usages primarily through market competition.
- Governmental policy: Government use of various forms of market auctions to reclaim bandwidth from less economically efficient uses of RF spectra.
- Reclamation of underused RF spectrum allocations: Government shifts of RF spectra historically reserved for public and military use to commercial users and reallocation of spectra among private and commercial users in response to market pressures.

- Globalization: Economic and market pressures on spectrum managers to reconsider compensation to national governments for use of RF spectra, the variations in economic value of RF spectrum across national borders, the ownership of RF spectra over the high seas and polar regions and in space, and the relevance of political boundaries to transmissions that cross borders by either wire or broadcast; clear prevailing trend toward globalization and away from nationalization as digital techniques, broadcast, and satellite telecommunications transcend government controls.

Both the Pelton merge and the Negroponte flip implicitly assume that wireless telecommunication will remain constrained to the RF spectrum. Neither projection includes the possibility of an entirely novel concept of wireless communications that is not limited to the RF spectrum. This revolutionary solution to a growing problem is the vision, the promise, and the possibility of laser satellite communications.

SOMETHING NEW: THE THIRD GENERATION

The RF spectrum is clearly a finite resource of increasing value. As demand for wireless services explodes, radio technology is persistently focusing on expanding the capacity of a small, preferred segment of the spectrum (3-30 GHz), because it meets so well the market demand for mobile—cellular—voice communications. It is likely that the powerful combination of technology and the market will be able to meet this demand well into the twenty-first century. To do so, however, lower-value communications—and those that do not require these preferred lower frequencies—must either migrate to higher RF bands, with all their inherent difficulties and costs; shift to fibre-optic cable, with the limitations and high costs of terrestrial and submarine construction; or follow a new path into a third-generation telecommunication system. The parallel market demand for high-speed, high-capacity communications —especially the Internet—is not constrained by the RF spectrum, since it does not emphasize mobile—or cellular—user capability. This demand for high capacity and high speed can be met by fibre-optic networks, but only within those networks. Although some countries are well connected through fibre-optic networks, global interconnectivity for most of the world depends on RF satellites. As usage of RF spectra and RF satellites shifts toward mobility and interactivity —personal communications services, cellular telephony, mobile radio, wireless networks, and direct broadcast satellite television—it is most likely that less valued applications will face increasing competitive pressures.

It is less likely that some combination of radio technology and market forces will be able to meet this demand for capacity, even by migration to Ka-band—or even to Q- or V-bands—in combination with the geographic limitations of fibre-optic cables. Something new is needed: a third generation satellite that is not constrained by the RF spectrum, that provides vast bandwidth with data rates measured in terabits/second (Tbps), and offers real cost advantages over not only the inefficient first-generation wire and radio technologies, but also the secondgeneration expanded-spectra RF satellites.

8

The Human Telecommunication

The invention of the telegraph and the telephone launched the human telecommunication enterprise with a single medium—copper wire. The invention of radio brought an alternative wireless medium, and expanded the possibilities for communication to the radio frequency spectrum. The penetration of these technologies into the market and establishment of infrastructures and institutions extended the enterprise into space. Fibre optics brought a new level of information-carrying capacity. The maturing second-generation human telecommunication enterprise refined the basic telecommunication media and technologies—copper-wire, RF wireless, and optical fibre—to improve performance, reduce cost, and expand connectivity.

As applications and users discovered their preferred media and channels, the pressures of the market encouraged migration to optimal RF bands. Operators have learned to use sophisticated science, new knowledge, and innovative techniques to obviate—not simply to solve, avoid, or fix—the problems and deficiencies of the first generation. As human telecommunications evolves into a mosaic of mature specialties, each providing optimal performance to a particular type of users and applications, both users and operators are stumbling on the dual problems of competition and interface. Specialization, migration, and optimization seemed to prevent telecommunication across media, notably between wire and wireless; between different types of users—computers and people; and between applications—voice and data. Although secondgeneration telecommunications are effective and efficient at the various discrete microlevels, they are often dysfunctional at a global macrolevel. The focus of the third generation is the global macrolevel; the challenges are capacity, congestion, and connectivity. Rather than simply wire and wireless, long-distance and local, telephony and Internet, or even broadcast and point-topoint, the third generation envisions multiple complementary, interconnected, noncompetitive communication specialties, each optimized to its particular function, physics, technology, and market. The third generation envisions holistic, hybrid, seamless communication and telecommunication across the entire electromagnetic spectrum, from DC to light.

FIRST GENERATION

Telecommunication's first generation was dominated by terrestrial wire and cable telephone and telegraph networks—first bare copper, then twisted pairs and coaxial, and most recently, single-mode fibre-optic cable. First-generation fibreoptic systems appeared in the late 1970s with speeds of 44.7 Mbps. Only in the last half-century has radio become an indispensable complement to terrestrial and submarine fibre-optic cable networks, first in the forms of microwave and tropospheric scatter links, and later as satellite links. Radio has not replaced wire; instead people learned to use radio and wire as complements in mixed-channel telecommunication. These first-generation mixed-media networks between wire and wireless technologies have become critically important and pervasive. First generation communication satellites in the geosynchronous earth orbit (GEO) have been providing fixed satellite services (FSS) at radio frequencies (RF) only since the mid-1960s. The critical capability of these first-generation "repeater" satellites is the ability to receive, amplify, and retransmit RF signals, either as a broadcast or as a single narrow beam. Most users—commercial, military, or personal—do not often need more than omnipresent repeaters above them.

Technological improvement of first-generation satellites, consequently, stresses higher power and greater bandwidth. The most modern, sophisticated of these large satellites transmit stronger signals through more transponders at higher frequencies, but do not offer more or better capabilities or applications. [1] Because, however, of the delay—latency—of up to a half-second while the RF signal travels up to the satellite in earth orbit at 36,000 kilometers (km) and back to the earth receiver, fibre-optic networks or single fibre-optic cables are preferable to GEO satellites for carrying voice telephony. Each first-generation satellite was an assembly of individual subsystems built to a design and specification unique to each mission. The huge costs and high risks of these early telecommunication satellites were borne largely by governments and military establishments—primarily the U.S. Department of Defence. "Over the past five decades, the [U.S. Department of Defence] served as the catalyst driving the initial development of space.

Taking a lesson from the airplane decades before, commercial exploitation followed the [military] use of the airplane. Essentially, the [U.S. Department of Defence] worked the problems out, developed the operational concepts for its use, and demonstrated its utility. We can clearly note similar developments regarding some space systems today." Reflecting the inexorable shift of space pioneering from government to industry, these large, heavy, expensive first-generation satellites still carry about a third of the voice telephony between countries, virtually all of the video telephony, and a small amount of data telephony. First-generation GEO RF satellites carry most of the voice, video, and data telephony to and from countries that do not have access to fibre-optic cables. In the 1970s and 1980s, the huge capacities of the new fibre-optic cables, relative to first-generation wireless technologies—AM broadcast, FM broadcast, microwave relay, or tropospheric scatter—and fibre's ability to transmit digital information instead of analog waveforms, quickly limited wireless to broadcast systems, where cables could not compete commercially.

A second niche reserved to wireless transmissions was —and remains—international telecommunications, where the costs of laying submarine cable were simply too large to be supported by commercial markets. In the late 1990s, most commercial satellites are used for telecommunications and carry signals in the preferred C- and Ku-bands. First-generation telecommunication satellites have traditionally operated in the RF spectrum in the L-, S-, C-, X-, and Ku-band frequencies, although some of the proposed LEO systems are to operate in P-band or Ka-band. [3] The most congested band is Cband allocated to geosynchronous satellite operations. [4] Of more than 200 current telecommunication satellites operating continuously in geosynchronous orbit 48 per cent carry video signals, 24 per cent carry voice, 8 per cent carry data, 11 per cent are used for earth observation, and 9 per cent are used for science. All operate transponders within the RF spectrum between 300 megahertz (MHz) and 300 gigahertz (GHz); none of the various systems include routine use of intersatellite links (ISL).

SECOND GENERATION

As terrestrial telephony matured, the telecommunication industry installed a vast infrastructure of twisted-pair copper telephone lines: telegraphy over single-strand wire virtually disappeared, as fibre-optic cables spread across the industrial countries. Superheterodyne FM radio forced AM radio into a specialized niche best suited to its lower frequencies. In response to demand for high-capacity circuits for "trunk and backhaul operations, carrying large chunks of transmissions from one place to another" fibre-optics network operators introduced multimode cables with higher capacities, speeds, and performance —45-Mbps T-3 links, 155-Mbps OC-3 connections, and 622-Mbps OC-12 cables. [5] "The advent of fibre-optics, digital data compression, and satellite relay hastened the global development of a comprehensive telecommunication system while driving down prices to the consumer of these services: international telephone calls are now infinitely more reliable, of better quality, and much cheaper per minute of conversation than in the past." [6] Optical amplifiers and high reliability are forcing twisted-pair copper wires into a specialized local niche market. As telecommunication systems and media mature in the second generation, operators optimize older technologies for specialized applications in stable markets, while developing innovative technologies specifically optimized for other—perhaps growing—markets.

The fashionable popularity of cellular telephones and laptop computers with digital data processing techniques provides wireless access to global information systems, private local area networks (LAN), and the Internet, as well as to the mature public switched telephone networks (PSTN). By increasing individual flexibility—through telecommuting or teleworking—digital telecommunication increases personal, corporate, and even national productivities. With a computer, a telephone, a telephone line, a modem, and some sophisticated software, work that can be done in London or New York can be done as easily and effectively —and at less expense—in Bangalore, Prague, Sydney, or Caracas. The cost of this geometric increase in human productivity is congestion across the first- and second-generation wires and cables that interconnect humanity's telephones and computers. Although second-generation technology has markedly increased the efficiency with which humans can transmit information to one another, it remains constrained by the physical capacity of the wires and cables that form terrestrial networks and of the radio frequencies that carry telecommunication traffic through space. Cellular telephony is a desperate attempt to circumvent the congestion of the world's wirelines—including the Internet backbone—with a shortsighted increase in the congestion of the world's RF spectra. The 1990s shift to cellular seems an ironic inversion of the 1970s shift to cable as humans continue their pursuit of bandwidth. Second generation communication satellites—notably the nascent mobile satellite systems—are generally built around a common spacecraft bus and modular designs for a range of related or similar missions.

Each bus contains a (1) modular power system, (2) command and data processing module, (3) attitude determination and control system, and (4) payload interface. Only the payload instruments differ between missions. Employing the same spacecraft bus for different missions greatly reduces the cost of satellite telecommunications"at the expense of not providing the most optimal design for a single given satellite mission." The larger, more powerful, new first- and second-generation FSS satellites are providing greater transmission powers, severe complexity, sophisticated software, and high gain antenna diameters to support a continually growing number of users. Second-generation MSS operators see a tremendous market for low-performance communication services—basically omnipresent repeaters. Omnipresence of simple communication capability is emerging the business paradigm of the leading commercial systems—Teledesic and Iridium. Designed to support cellular telephony, the new second-generation satellites occupy lower orbits to reduce latency and converting conventional analog communications signals to streams of digital bits before transmission. Operating in low earth orbit (LEO), midlevel earth orbit (MEO), or highly elliptical orbit (HEO), constellations of these modern, light, small, specialized second-generation RF communication satellites will provide mobile satellite services (MSS) at radio frequencies that are both higher (Ka-band) and lower (VHF and UHF) than those associated with first-generation FSS.

The voice-oriented systems "will provide cellular telephone service via satellite.... [T]hese new personal communications satellite systems will be based on a relatively large number of satellites in orbits considerably lower than geosynchronous ones." Other second-generation satellite systems carry data and video, especially through Internet connections. Not necessarily competitive with either mature FSS systems or with fibreoptic networks, the new second-generation personal communication service (PCS) systems provide the respective advantages of cellular telephone systems and satellite carrier systems in a single global network. With initially less capacity than either FSS or fibre, these new MSS satellites will operate, in effect, as orbiting cellular base stations with which mobile transceivers or handheld cellular telephones will communicate directly. Like the general-purpose first-generation FSS systems, the specialized second-generation MSS systems will be capable of serving large areas, mobile receivers or personal handsets, and remote sites where fibre-optic networks (FON) or fibre-optic cables (FOC) are not available. Perhaps the most significant change in patterns of human telecommunication inherent in the second-generation MSS systems is that "space-basedsystems will increasingly deliver information directly to the consumer, rather than to a commercial data hub." [13] The second-generation PCS systems are changing the lives of much of humanity, providing rapid access to information through handset terminals across the planet. Even as the torrent of communication becomes a flood of information overload, demands for both capacity and speed and for convenience and mobility are unlikely to abate. "The pressure on spectrum will intensify, because high data-rate signals need much more 'bandwidth'—a far bigger slice of spectrum-than do low data rates."

CONGESTION, MIGRATION, AND SPECIALIZATION

First-generation FSS satellites—like first-generation terrestrial networks-carried whatever traffic that appeared, regardless of efficiency, suitability, value, or even market demand. Both satellites and cables carried data and video as if they were voice. Far more sophisticated second-generation MSS satellites focus on a specific primary application—duplex voice for mobile, cellular terrestrial transceivers—while also providing services for which they are less well suited —data and video telephony. Like first-generation satellites, traditional terrestrial circuit-switched wire-cable networks accept all traffic and treat each signal as voice. Second-generation packet-switched networks—wireless cellular, wireline fibre, and even MSS-optic—treat all traffic as data and must reconvert any voice signals back to voice for delivery.

Although market forces and technological ingenuity continue to introduce improvements, one result of using both wire and wireless for all sorts of traffic has been hopeless congestion of both preferred RF bands of spectrum and fibreoptic networks. The second-generation telecommunication infrastructure is becoming a jumble of competing, alternate, overlapping, often incompatible, and usually poorly integrated technologies, each with its own high-tech device, address, and set of transmission channels.

With the efflorescence of devices and applications, the growing population of service providers continues to demand ever more transmission capacity—and usually also high speed. At least five separate types of service provider are contending for bandwidth in the RF spectrum: telephone companies; cable television systems; terrestrial and satellite RF broadcasters; Internet access and service providers; and on-line service providers. Another consequence of the advent of cellular systems is intense competition for the scarce resource of transmission capacity appropriate—in frequency, wavelength, speed, and modulation—to the application and to the market to be served. The second generation of telecommunications is shifting toward competition through specialization. The RF spectrum and its capacity to carry information are absolutely finite, and many applications require a particular band of frequencies —cellular telephones are most useful at frequencies below Ku-band. Each medium—RF broadcast or fibre-optic cable—is optimal for some applications and less suitable for others, but some applications are more demanded by users.

Although innovative radio and optical technologies—code-division multiple access (CDMA), time-division multiple access (TDMA), frequency reuse, beam polarization, wavelength-division multiplexing (WDM), and dense wavelengthdivision multiplexing (DWDM)—clever software, and complex protocols increase the efficiency of spectrum usage, they neither obviate nor resolve the problems of insufficient transmission capacity at preferred, optimal frequencies. For maritime, aeronautical, military, and broadcast communication services, where terminals are mobile, but need not be personal, cellular, or handheld, the mature technologies of first-generation FSS systems remain most suitable even for future enhanced systems. The increasing consumer demand for PCS systems that permit telephone calls through wireless handsets to reach destinations anywhere on earth, however, is best met by constellations of satellites in LEO or MEO. Low latency and attenuation—15-20 decibels—lower power requirements, and small, inexpensive ground receivers augur continuing growth of this cellular application and market. Requirements for high-speed, high-capacity, point-to-point, long-distance telecommunication are best met by fibre-optic cable, if it can be practicably installed.

Innovative networking technologies and products to replace modemenabled first-generation telephone connections are shifting Internet traffic to Internet-specific networks. "Internet access *via* the LAN or WAN [wide area network] is definitely going to facilitate a better quality of service for your higher-end applications, because the bandwidth is going to be there. You don't have to fight with a lot of other users to gain access to that as you do with dialup service." It is neither realistic nor cost-effective, however, to envisage developing countries—or even prosperous advanced countries with sparsely populated areas—establishing high-speed, high-capacity fibre-optic cable networks across their entire territories. Instead, in addition to fibre-optic networks where they are cost-effective, operators are learning to connect important regional communication nodes—probably embedded in copper twisted-pair or coaxial cable local networks—through RF or laser FSS links—to each other and to the rest of the world. Local spot beams from LEO-MEO satellites are expected to provide cellular coverage where demand is greatest and where users can absorb the projected high unit costs. In poorly served remote or rural areas, wireless local loop technologies, which are emerging in urban areas as a low-cost solution to the "last mile" problem, may be linked through microwave or fibre-optic links to GEO satellites for regional connectivity. Some innovative entrepreneurs are proposing direct wireless connectivity between wireless loop systems and MEOLEO systems for local cellular connection, and possibly—in theory—through laser satellites for global access.

Indeed, wherever terrestrial networks are not commercially competitive—because of low traffic density, high capital investment cost, inadequate infrastructure, or something else—or not suitable —maritime, mobile, and aeronautical environments—the complementarity between specialized RF satellite communication systems and narrow-purpose terrestrial networks is particularly powerful. Current second-generation infrastructures and technologies can route voice or data telephony around the world, broadcast to every inhabited place on earth, and offer the prospects of global cellular telephone conversations and connectivity through the Internet for both computers and people. Neither specialization nor technological innovation, however, have been nearly as effective in relieving the congestion that prevents universal global connectivity. To obtain the capacity needed at the preferred frequencies for this vision requires less-than-optimal applications to migrate to higher frequencies and shorter wavelengths beyond the RF spectrum. Not every application can migrate: cellular telephones must operate inside buildings and in cities where only long wavelengths are effective. Other applications that do not require long wavelengths can migrate to higher frequencies and decongest the bands that cellular telephony requires. Although third-generation laser satellite telecommunication cannot provide universal connectivity, it does allow the migration and specialization that can relieve the congestion in UHF, EHF, L-, P-, C-, and Ku-bands that is preventing it.

THIRD GENERATION

Dealing with the persistent RF bandwidth problem requires an imaginative third generation of telecommunication satellites whose development must begin even before the second generation is fully deployed. The new third-generation satellites portend a reversal of the trend of continuing growth in satellite size and weight from the early 1950s. Whereas first- and second-generation commercial satellites continue to grow in size, complexity, and cost, the sizes of the new third-generation satellites began to decrease in about 1990-1991. [20] For the third generation of the telecommunication enterprise, the centre of power is not in the creation of ever-larger, more powerful, supersophisticated, individual RF satellites, but in the integration of many small laser microsatellites into a more powerful communications constellation. One laser satellite constellation replaces hundreds of RF satellites in providing high-capacity global connectivity.

Built around the low-cost/high-risk "smallsat" design strategy, thirdgeneration laser satellites involve multifunctional structures (MFS)—integrated electronic, thermal, mechanical, and structural functions—nanotechnology, microelectronic mechanical systems (MEMS), and distributed satellite systems —constellations of many satellites working together to accomplish a mission. Seeking minimal costs, some smallsats, particularly those employing mature, proven technologies, also eliminate redundancy—the single-string design—and simply accept a moderate risk of single-point failure. "Instead of developing satellites weighing thousands of kilograms and costing hundreds of millions of dollars, engineering teams of only a handful of people began designing 'smallsats' weighing 200 kg or less and costing only a couple million dollars. The size of these small satellites also reduces operational costs, for now the satellite may be launched on a $9 million Pegasus rather than a $78 million Atlas class rocket." Also called nanosatellites or microsatellites, these third-generation smallsats are being deployed—as production prototypes—in powerful constellations with capabilities far beyond those of first- and second-generation systems. Two significant advantages of these small, simple microsatellites are their low production costs and their low weights. Especially significant to commercial telecommunication satellite operators, lower costs and weights of third-generation satellites arise primarily from simpler and less satellite hardware, but also from the small, aggressive manufacturing companies that make the new thirdgeneration microsatellites.

These companies work closely with operators to develop telecommunication constellations and systems instead of simply selling satellites to commercial or military operators. Particularly well suited to laser telecommunication applications, microsatellites are also apposite to remote sensor imaging and astrophysics experiments. The primary benefit of the third generation, however, lies less in technology or cost than in the dramatically increased utility of "lots of little" rather than "a few big" satellites. Most first- or second-generation satellites operate either as remotely or automatically controlled individual sensors that download data to ground stations or as communication echoes, which repeat what they receive from the earth. In contrast, the laser satellite communications constellation is analogous to a distributed, remote, automatic, autonomous robot. The performance of the laser satellite constellation—or any third-generation satellite constellation—is more than the capabilities and scope of individual satellites performing programmed tasks. Each satellite is no longer important in and of itself, and does not need dedicated control systems. Neither, however, can any single satellite accomplish the mission of the constellation. These third-generation laser satellites eliminate the RF spectrum congestion by providing laser communication capacities beyond those available in any RF spectrum bands. Third-generation laser satellites will provide transmission speeds at least 1,000 times faster than those of the second-generation RF satellites. Rather than data rates of megabits per second (Mbps), third-generation communicators will demand—and receive—terabits per second (Tbps).

The physical limits of the RF spectrum and radio science and antenna technology seem unlikely to be able to accommodate these demands, at any price. To achieve these data rates, wireless communication will need to abandon the RF spectrum and migrate applications with high capacity requirements to the optical spectrum to operate at terahertz, rather than gigahertz. Just as transoceanic telephony migrated from copper wire to fibre-optic cables, high-capacity, highspeed atmospheric and space telecommunications must migrate to shorter wavelengths with their inherent higher capacity and speed. The carriers in these optical spectra are laser beams, not radio waves. Discovery of the ability to modulate light and to send light beams through terrestrial and transoceanic fibre-optic cables expanded the human telecommunication enterprise beyond the relatively small capacity of the RF spectrum into the vast possibilities of the optical spectra. The more recent discoveries that humans could use modulated laser beams to carry and transmit information through space has completed the syllogism, shown the feasibility of laser communications, and opened to mankind virtually unlimited possibilities for communicating: the third generation.

Laser satellite communication is a vision of the third generation of human telecommunications, of satellites linked to terrestrial networks, and of lasers linked seamlessly to fibre-optic systems. The third generation—laser satellite communications—is a realization of the optimistic notion of a holistic telecommunication system, an integrated, seamless flow of humanity's messages through the best media and at the best frequencies for the sender, the traffic, the user, and the application. The introduction of laser satellite communication to humanity's telecommunication enterprise allows, invites, and encourages the migration of traffic that requires very high capacity and speed to higher frequency bands—ultimately to free-space laser and fibreoptics. It stimulates rational, deliberate specialization of applications for those frequency bands—RF or optical—wire or wireless channels, and markets best suited to them. Both radio and cable operators may specialize in those applications that require a particular carrier or frequency—probably the most demanded and profitable—and transfer less critical applications to higher frequencies.

The several types of wide-band traffic can be expected to migrate to higher frequencies—ultimately laser beams—release capacity at lower frequencies that are preferred or required for other applications, and stimulate efficient, effective, profitable third-generation specialization. Instead of political, arbitrary—or even market-based—priorities that preempt less urgent or less profitable messages, the holistic system can route voice traffic through the low-frequency RF broadcast and point-to-point channels, fibre-optic cables, or copper wire lines that best suit it. The system would shunt video and television through higher RF channels or high performance fibre-optic networks, and transfer traffic carrying very high volumes of information —Internet, multimedia, imagery, or data—to the very high optical frequencies that can carry its huge volumes at very high speeds—laser beams. summarizes the evolution of the telecommunication enterprise from copper wire and radio to the third generation of fibre-optics and laser.

COMPARING WIRE, RADIO, FIBRE OPTICS, AND LASER

With drastically different physical characteristics, market demands, and preferred applications, systems using copper wire, RF satellites, fibre-optic cables, and laser satellites need not be competitive. Radio is inherently preferable for multireceiver broadcast; copper wire, fibre-optic cables, and laser beams can reach multiple users only through networks. Wire and fibre use physical conduits that must be installed within expensive infrastructures. Copper wire and RF are limited to frequencies below 100 GHz with correspondingly low transmission rates and capacities. Optical fibre and laser beams carry modulated light at frequencies into the THz range. Each medium—metal wire, RF, fibre optics, laser—has an optimal application, a specialty, a niche in the telecommunication enterprise. They cannot be compared directly in all respects, but only with reference to specific applications and the ultimate cost to the final user.

Copper Wire

In 1844, Samuel Morse sent the first telegraph message over copper wire from Washington, D.C. to Baltimore. For Americans, the era of national telecommunication began in October 1861 with the completion of transcontinental telegraph lines. The earliest telephone systems provided voice communication between telephones in buildings, usually businesses, and residential telephones did not become common, even in the United States, until the 1940s. Today, nearly all local communication and much long-distance telephony still pass through copper wires, cables, switches, routers, and the networks that connect local lines together.

The basic telephone line in common use across the world in 1999 is fundamentally the same line that was used in 1940: twisted-pair copper strands. Category 5 copper cable—four twisted pairs of copper wire—still used heavily throughout the world, has a maximum operating frequency of 100 MHz and a distance limit of 100 meters without degrading a voice signal. Enhanced category 5 and category 6 cables are rated at 200 MHz, and category 7 (nonstandard and not certifiable) operates at 600 MHz. Copper cables and networks suffer from a number of disabilities: circuit crosstalk, near-end crosstalk, RF interference, electromagnetic interference, line attenuation, skew, propagation delay—latency—return loss, impedance mismatch, as well as low capacity. In addition to the wire itself, however, the connectors, jacks, patch panels, switches between strands, and any splices or repairs, are sensitive to mismatches, anomalies, or imperfections.

Radio

James Maxwell predicted electromagnetic waves in 1864, and Heinrich Hertz elegantly verified Maxwell's predictions in the 1880s. Guglielmo Marconi introduced mankind to the RF spectrum with his "system of wireless telegraphy able to transmit intelligible Morse signals to a distance of over ten miles." W.J. Clarke designed the first wireless telegraphy apparatus that the United States Electrical Supply Company introduced as the first commercial radio device. In 1901 Marconi's equipment sent coded "short wave" signals from England to Newfoundland. Lee De Forest's invention of the vacuum tube amplifier in 1908 made wireless telecommunications practicable. Since Marconi continued his experiments and research to discover ionospheric reflection, scientists and engineers have continued to refine mankind's knowledge of the RF spectrum. The first commercial uses of radio were probably the radiotelephones installed in many oceangoing vessels as early as the First World War. Simplex two-way mobile radio systems appeared ashore in the 1920s and developed into full-duplex "diplexed" systems in the 1940s. Although AT and T engineers proposed the concept in the 1940s as a "high-density mobile system," the first cellular telephone system appeared in Tokyo in 1979.

The concept was introduced in the United States only in the 1980s. Different wavelengths and frequencies were found best suited to particular purposes, based initially on discoveries that various things—other radio signals, foliage (above 1 GHz), buildings, hills, rain, oxygen molecules, fog (above 300 GHz), sun spots, and cosmic rays—caused increasing interference, blockage, and attenuation. The market demand for more speed, more services, and more capacity pushed frequency usage into ever higher spectral bands. Despite their inherent difficulties with precipitation attenuation and atmospheric scintillation, higher frequencies afford broader spectra for useful exploitation. As a practical compromise between capacity and quality, terrestrial telecommunication usage focuses in bands below about 1,000 MHz, satellite systems operate up to about 75 GHz, and microwave and millimeter wave systems operate up to 300 GHz. Although shorter RF waves occur above 300 GHz, telecommunication usage jumps from millimeter wave to infrared and lightwave bands, which are currently used for only a few specialized applications, despite their truly enormous capacities and virtually infinite transmission capacities.

Since light and infrared signals are subject to environmental interference in atmospheric free space (wireless), these media have historically been used only for very short distances (a few hundred meters, or between buildings) or in controlled atmospheres. Some salient applications at these high frequencies include remote controllers for computers and television sets, wireless LAN connections, and linkages between computers and their peripheral equipment. In some applications at these frequencies, infrared and light signals are sent through fibre-optic cables.

Fibre Optics

Fibre-optic cables are either plastic or glass. In contrast to the Mbps speeds of plastic—or dark—fibre, speeds greater than 2.5 Gbps are possible through the bright glass fibre. Glass fibre retains full carrier integrity in runs up to 300 meters and retains motive force at distances of hundreds of miles. Plastic fibres show inherently greater attenuation—intermodal dispersion—than do glass fibres, with lower data rates. Like copper, fibre-optic cables involve signal attenuation, although this is much less constraining in glass than in copper.

Connectors between fibre cables are even more sensitive and vulnerable to imperfections than are copper connectors. Although no universally accepted standards for fibre-optic connectors exist, vendors are introducing fibre-optic connectors to complement their lines of copper connectors. Fibre-optic cables —both single-mode and multimode fibres—are also vulnerable to signal dispersion or attenuation, which imposes distance limits on fibre-optic transmission and the need for properly spaced amplifiers. Multimode fibre can support gigabit Ethernet or higher frequency technologies for up to 100 meters. Since fibre achieves much higher bit rates over longer distances and is less affected by electromagnetic interference than copper, "for very high speeds, capacities, and longer distances, multimode fibre is preferred." Fibre optics has been historically applicable to five communications applications, or markets:

- Interoffice trunking—heavy-usage lines between central offices or users; operating at low rates (< 1,000 mbs) over up to 80,000 channels, rarely more than 50 kin in length; fixed satellite services are not competitive in this sector;
- Local loop distribution-movement of concentrated traffic to receiver sites: operating at low rates (< 1,000

In 1998, the U.S. National Reconnaissance Office (NRO) awarded contracts to TRW to design, build, and operate the Geosynchronous Lightweight Technology Experiment (GeoLITE) satellite with laser communications experiments. Parallel and complementary U.S. Air Force research in search of laser antimissile weapons has developed several effective techniques for overcoming atmospheric attenuation of laser beams. "Every photon—not electrons—in a laser beam travels in exactly the same direction and has the same frequency and phase of oscillation. This property makes laser light very easy to control precisely.... Laser beams are noninteracting—they can cross without affecting one another at all." These features —coherence and noninterference—are direct converses to the two major technical inconveniences of RF and lead to the utility of a modulated laser beam for telecommunications. Ironically, they also hold the key to overcoming the only limitation on laser satellite communication: the susceptibility of laser beams to atmospheric disturbance.

COMPARATIVE ADVANTAGES

A number of comparisons may be made along various criteria, with widely divergent conclusions that may be critical or irrelevant to a particular application. Copper-wire networks remain price competitive—in some places the economically preferred medium—in local, low-volume, low-speed, voice telephony service. It is, thus, unlikely that copper wire will be completely replaced in the final links to individual telephone users. Radio satellite systems are clearly a preferred medium for global broadcast telecommunications, mobile and cellular applications, and multipoint broadcast networks. Fibre-optic systems are essential channels for heavy, terrestrial, domestic, point-to-point traffic demanding high speed and high capacity. Laser satellite systems provide complements and supplements to these other media and are especially effective in carrying heavy global point-to-point traffic requiring very high speed and very high capacity. One useful, important comparison involves the latency of the various systems. For simplex broadcast, radio, television, and point-to-point data transmissions, latency is acceptable.

Human users of duplex voice circuits find it at best annoying, and generally unacceptable, whereas computers are often completely unable to deal with it and simply reject the signal. When the one-way time delay exceeds about 0.3 second—interaction latency of 0.6 second—human speakers experience confusion; time delays of less than 0.2 second are imperceptible to humans in telephonic speech. Computers and digital processors, communicating at hundreds of Mbps, cannot tolerate interactive delays of more than a few milliseconds. The primary sources of two-way latency in domestic telephone networks —copper wire and fibre-optic cable—of 0.010-0.040 second lie in the transmission and switching architectures. Even in such large countries as Russia, the United States, China, Brazil, or India, these circuit delays rarely exceed the 0.2second human latency threshold. The periodic buffers and booster relays in a transoceanic cable raise latency to 0.200 second. The round-trip propagation delay for LEO satellites is 0.005-0.010 second, and the MEO increases latency to about 0.070-0.080 second. Although the distance to GEO imposes a latency of 0.250-0.270 second, any additional increments of terrestrial network switching, voice code processing, intersatellite linking, or multihop relays make GEO connections generally unacceptable for duplex voice communication. "Time delays have, in fact, virtually eliminated geosynchronous relay for voice communications *within* the borders of the United States and western Europe. However, roughly 70 percent of international continent-to-continent message traffic still streams through geosynchronous satellites, but even in that case, underseas fibre-optic cables have been nibbling away at market share." compares typical latencies of terrestrial and satellite telecommunication systems.

Table: Comparative System Latency (Thousandths of a Second)

	PSTN Telep-hone	**LEO Sat -ellites**	**MEO Satellites**	**GEO Satel-lites**
Propagation		5-10	70-85	250-270
Multiplexing, TDMA		320-360	320-360	320-360
Single hop link				360-730

Multiple intersatellite link		860-890	400-490	1,080-1,300
Buffers, relays, amplifiers	100		270-320	270-320
Digital compression andarchitecture	5-20	50-80	50-80	50-80

In addition to propagation delays, computer processing associated with digital compression adds an increment of 0.05-0.08 second and the groundstation circuit-switching used by many MSS systems can take another 0.07 second. TDMA systems routinely store data packets in buffers for up to 0.10 second while signals are in transit. Intersatellite links may take as much as 0.09 seconds for each buffer's processing. With these levels of latencies, most telephony carried over LEO satellites will be fully acceptable to humans, whereas MEO and GEO satellite systems will probably present notable, or unacceptable latencies, especially for international calls. Constrained by the low data rates of all but the fastest fibre-optic cables, computers operating over terrestrial networks can tolerate their relatively low latencies.

The low data rates of first- and second-generation RF satellite frequencies—up to 30 GHz—similarly prevent computers from operating at speeds that would not tolerate the latencies of LEO and MEO. The very high latencies of the GEO, especially for a digital signal through the PSTN to an intercontinental recipient, however, hinder interactive computer traffic at all but the lowest operating rates. Another useful comparison may be made in terms of costs and prices of telecommunication services. Historically, the total distance of carriage has been a fashionable criterion for comparing wire and wireless in a specific situation. The costs of purchase, installation, and maintenance of wire or cable are directly related to its length, whereas the cost of a radio wave—or laser beam—is unrelated to distance, except through power requirements.

The differing sensitivities of the technologies to distance, transmission intensity and volume, number of points served, capital investment, and operating costs make any price analysis explicitly subject to a specific application. Although cellular telephony has a relatively high cost per unit of transmitted information, short range, and low power, it can serve an immense number of users and locations, requires minimal capital investment, and is easily connected to any of the long-distance —or global—networks; its primary advantage is probably its mobility, convenience, and flexibility. "Fibre has bandwidth capabilities that dwarf anything that cellular can provide and the cost per bit is almost nothing. But it's fixed [not mobile]." The major technological developmental trends in each technology have prosaic effects on pricing: costs fall steeply toward some lower limit as the technology matures, but are followed by a gradual decrease in gains at the cost of higher investment. Since wire technology is the most mature, its costs have fallen most significantly and seem to be stable and relatively low. RF technology initially generated steep economic gains with increases in radiated power and later with use of higher frequencies. The combination of high power, high frequency, and huge coverage from the GEO quickly recovered the high capital costs of satellites, and became a new source of high-profit revenues.

Even greater profits from satellite systems emerged as ground stations were simplified into simple portable antenna systems and the development of small aperture antenna systems brought additional users into the market. After recovering the initial investment in a satellite, the RF satellite industry has only minor infrastructural costs, especially for broadcast applications.

Although already sophisticated, fibre-optic technology is still developing and can be expected to experience further—probably small—reductions in unit costs. The maturity of fibre-optic technology and the cable industry have created a substantial infrastructure—like the copper-wire infrastructure—that can be extended by adding relatively inexpensive terminal and interface structures. Whereas installation costs of fibre-optic cables are the primary investment in these systems, the actual cost of materials—fibre-optic cable—is a marginal element of investment, which is dominated by labor, right-of-way, repeaters, signal generators, amplifiers, circuit multipliers, input-output devices, and terminal systems. Increase in bits per second (bps) capacity, however, requires expensive replacement of obsolete cabling and equipment, in addition to adoption of new technologies and techniques—asymmetric digital subscriber lines (ADSL) or local multipoint distribution systems (LMDS). Like terrestrial laser communication systems, laser satellite communication is still an emerging technology and has not yet felt the cost reductions inherent in standardization, mass production, economies of scale, or market penetration.

With no requirement for a vast, dedicated infrastructure, laser satellite communication requires smaller investment than a global fibre-optic network. With no need for large constellations of LEO-MEO satellites or large, powerful GEO satellites, laser satellite communication requires less investment in the space segment than other satellite systems. Although the signal segment relies on technologies that have not been commercialized, virtually all components and devices are already available in the commercial market. The major investment for laser satellite communication is the few sophisticated ground stations to manage the network and connect to terrestrial systems.

APPLICATIONS AND SPECIALIZATION

The physical, intrinsic characteristics of copper wire, RF satellites, fibreoptic cable, and laser satellites determine specific applications and markets for which each cannot compete. Wired networks are static, easy to control, and do not radiate energy into the environment. Wireless systems are dynamic, difficult to control, and always radiate energy to interfere with other signals or endanger humans. Hierarchical wire or cable networks funnel traffic through the relays, amplifiers, switches, routers, and nodes that form most of the world's telecommunication infrastructure. Wireless RF is most efficient in matrix networks that bypass traditional interconnection nodes. Four factors—distance, capacity in bps, number of points served, and regulatory concerns—identify markets or applications for which one medium has a decisive competitive advantage. The primary advantage of wireless RF broadcast lies in its instant ability to connect virtually any point to any other point: universal interconnectivity. Wire, fibre, and laser cannot do this, nor can wire or fibre-optic systems be instantly reconfigured to add or subtract points of service. Wire, cable, and laser are excluded from applications involving mobile receivers or transmitters, because of their point-to-point limitations; mobile communications is a large, growing market for RF satellite carriers. This fundamental systemic difference between wireless and wired service leads to quite different regulatory concerns.

The increasingly salient issue of the hazards of electromagnetic radiation creates a culturally sensitive obstacle to expansion of wireless systems, especially at high frequencies. The parallel problem of radiation interference with other signals is equally perplexing. Wire and cable systems simply do not have either problem. Very short-range, low-capacity links within a company, town, or compact location are best provided by wire or cables, with laser often preferable for veryhigh-capacity traffic. Fibre-optic cables are the preferred medium for midintensity traffic between two distant fixed points, although advantage over satellite declines with distance. Although existing telephone networks can easily provide capacity for all of the world's voice telephony, market demands for data and video capacity are saturating those systems. Extending the three core services of the multimedia age—voice telephony, cable television, and Internet access—to global interconnectivity is placing demands for capacity on wire and cable networks that they cannot meet. RF satellites have a clear advantage in distributing low- and medium-intensity traffic among multiple receivers and multiple transmitters.

The introduction of very small aperture terminals (VSAT) and direct broadcast satellites (DBS) explicitly exploited the RF satellite's advantage in providing high effective isotropic radiated power (up to 200 watts) and narrow RF beams (focused to 1°). In the converging markets for longdistance point-to-point cable television, voice telephony, and Internet access, the technologies may be competitive, alternative, or complementary. The continuing, growing importance of converged sector requires increasing aggregate carriage capacity beyond that available in the RF spectrum, alternative transmission possibilities to increase reliability and availability, and options for reducing the price of transmission for low-priority traffic.

Beyond the major area of direct competition among the four carriers in the converged multimedia sector, each technology offers possibilities for dominating the fields for which it is uniquely suited and for adapting to other fields in which it may be less than the preferred medium, but merely an alternative. As the market for telecommunication services develops, the need to communicate between mobile platforms—whether handheld, cellular, or vehicle-mounted—at lower radio frequencies (no higher than C-band) is forcing long-distance point-topoint communications to migrate to higher frequencies and to fibre-optic carriers. Through specialization, interconnection, and complementarity among the four technologies, and among their respective networks and systems, the global telecommunication enterprise can develop secure and reliable connectivity, as well as efficient use of telecommunication resources where and how they are best employed. Although it remains unclear whether wired communication or wireless communications systems will eventually predominate, few would argue that integrated, hybrid, converged, or holistic communication systems will become typical—even standard—in twenty-first-century communications. Presents several useful, but general, comparisons across the four technologies.

9

Satellite Communications

As humanity moves into the third generation of telecommunications, it is also learning to deal with what many have come to know as the Knowledge Era, or the Information Age. As humans explore more deeply the knowledge economy, they are slowly realizing that it is business, not merely technology, vision, promise, or concepts that creates reality. At the centre of modern society lies not the dreamer, the politician, the scientist, or even the individual. At the centre of the economy, the community, the polity is the managed institution, including the modern business enterprise. It is the modern business that embodies and exploits technology, information, productivity, innovation, and even the individual. The modern business, and the governmental, religious, social, and other managed institutions that support it, are modern man's way of accomplishing things. If laser satellite communication is to achieve its promise, it must become a business. These pioneer LSC companies have much more to do than simply building a business or executing a business model. Theirs is the challenge of creating an industry that empowers all of humanity to leap into the next generation of the telecommunication enterprise.

TECHNICAL SYSTEM CONCEPT

The nucleus of a laser satellite communication (LSC) system is a set of six spin-stabilized satellites in midlevel earth orbit (MEO) above the Van Allen radiation belts. A smaller set of three to five laser satellites in geosynchronous earth orbit (GEO) expands coverage, extends connectivity of the system to higher latitudes, and provides additional capacity for trunk, simplex, and non-real-time traffic. As they move around the earth in a circular orbit, the six intervisible MEO satellites would be deployed in two three-satellite laser rings offset from each other in longitude, while the GEO satellites would be positioned to achieve global connectivity and coverage of target markets. Laser intersatellite links (ISL) would create a double laser ring at MEO linked to another GEO laser ring, both circling the earth. These interconnected laser rings would carry low-latency duplex and voice telecommunication traffic as modulated laser beams between the satellites. [1] Laser uplinks and downlinks with regional ground stations would establish laser channels for transmitting traffic to, through, and from the laser rings and between the laser rings and any ground stations.

Through such a dual GEO-MEO constellation of laser satellites, modulated laser beams can carry digital and analog traffic between two terrestrial points at rates of gigabits—and even terabits—per second. The use of modulated—amplitude, phase, or frequency—laser reflection technology for LSC reduces the weight and size of satellites and substantially reduces the criticality of satellite attitude control. The resulting laser microsatellites are small enough that each telecommunication constellation of laser satellites could be deployed by a single launch vehicle. Small satellites and boosters designed for responsiveness provide cost-effective capabilities for graceful system enhancement, replacement, or degradation. Binary failures of heavily invested first- and second-generation spacecraft and the ambiguous in-orbit failures arising from long-term effects of space weather degradation increase both the cost and the risk of orthodox telecommunication satellites. The long lead times of all heavy-payload launch schedules and the continuing arguments between launchon-need and launch-on-schedule complicate the simple replacement of an obsolete, degraded, or nonoperational satellite.

For the third-generation laser satellite constellation, issues of replacement are simple and less stressful. Since the loss of a single satellite represents only a small fraction of the system's capability, replacement of a satellite will *never* be urgent! Because of the flexibility in choice of launch vehicles, schedules, and orbits for laser microsatellites, replenishment can occur with the efficiency of deliberacy, rather than the "cost-plus" urgency inherent in the crisis of subsystem failure in an orthodox RF satellite. Many of the crippling disabilities of orthodox GEO satellites—orbital positioning conflicts, frequency interference, revisit delays, high costs and risks of spacecraft and launch, or latency—are largely negated by the powerful synergies of laser-linked MEO constellations.

The individual laser satellites comprise elements, subsystems, and technologies that are neither new nor different: cost and risk of subsystems and components are relatively low. Unlike the mega-LEO constellations, however—and in stark contrast to the local "footprint" of each first-generation FSS satellite—the laser satellite rings provide global connectivity with only a few simple microsatellites: complexity and sophistication remain in the ground segment. The novel, inverse architecture of a third-generation LSC system generates commercial might and technical promise in the responsive and pervasive presence of an entire system launched or stored on orbit, ready to communicate. "One satellite is cheap; many together are power." This same technology, however, also increases the satellite's capacity to transmit information. The high frequency, capacity, and speed of laser beams using smaller, lighter antenna systems (telescopes) with lower power requirements make them an attractive alternative to the high-power RF signals operating at lower data rates in congested frequency bands with lower capacity. The laser medium is optimally suited for use of wavelength multiplexing techniques to exploit the wide range of laser carriers for transferring multiple high-density messages simultaneously and efficiently.

PRINCIPLES

Any LSC system embodies the principle of modulating laser beams to carry information. The key to efficient communication over laser—rather than RF—lies in the extremely short wavelength of light—about a thousandth of that of RF. This difference allows the focusing antenna to be about a thousandth of the size and weight of the RF antenna and the data rate to be 1,000 times greater than the rate of RF transmissions. An optically communicating satellite can either be the size of a comparable RF satellite with 1,000 times the available capacity, or provide similar capacity in about a hundredth of the size of a comparable RF satellite. As with the well-known concepts of FSS communications, the LSC concept involves three basic elements:

- The space segment—design, features, control, and capabilities of the satellites; signal processing on board the satellite; means of launching them into orbit; the dynamics of the satellites' orbit; and the mechanics of positioning and attitude control;
- The signal segment—the carrier medium used for communicating to, between, and from satellites—uplink, intersatellite link, downlink; the effects of distance; sources of signal interference, attenuation, and instability; modulation schemes and protocols to ensure proper transmission and reception; effects of digitization and compression;
- The ground segment—placement, construction, functions, and operation of ground stations; antennae, receivers, transmitters, and other ancillary equipment; antenna engineering; multiplexing and multiple access schemes; tracking and telemetry control facilities; interface with terrestrial telecommunication networks and systems; administrative structure; managerial coordination of operations, finances, and business activities.

THE STATE OF THE ART

Since the mid- 1970s, both industry and governments, as well as scientists, have explored the possibilities of laser-carried communications between ground stations, aircraft, and satellites. One of the most promising U.S. experiments is a demonstration LSC system developed by AstroTerra Corp. for testing on the second Space Technology Research Vehicle (STRV-2) launched by the U.S. Ballistic Missile Development Organization in 1998. The AstroTerra system transmits 1.2 Gbps over a distance of 1,800 km, occupies less than 1,000 cubic inches, and weighs 14.3 kg. The satellite package comprises the transceiver telescopes and gimbals, an electronics module, and a deployment mechanism. The system employs single 5.4" Schmidt-Cassegrain receiver telescopes at both the satellite and the ground stations. Each of the eight communications lasers operates at 125 mW at a wavelength of 810 nanometers (nm) with a divergence of 80 microradians. Four of the communication lasers form one 600-Mbps channel with right-hand circular polarization, and four form the second channel with left-hand circular polarization. Depolarizers separate the communication signal into the two carrier channels for conversion—by two avalanche photodiodes—into an electrical digital signal for decoding. Power switching the laser diodes modulates the transmitted data onto the laser mean.

Two acquisition and tracking lasers—100 mW each at 852 nm with 500/1,500 microradian divergence—establish and maintain beam integrity. Cesium atomic line filters in the receiving telescopes separate the communication signal from the tracking beacon. The receiving telescopes focus the acquisition-and-tracking signal onto a camera sensor to track it and drive the gimbal system. The powers of modern laser diodes and the magnifications of commercial telescopes are well able to transmit and receive coherent laser beams at interplanetary distances. The subsystem with the greatest risk is, thus, the pointing system. Since the radial dispersion of the laser beam has a Gaussian distribution, to maximize the power received the receiving telescope must point within ± 30 per cent of the centre of the beam. Although the STRV-2 has a generous pointing error of ± 40 microradians, high precision bearings, sophisticated resolvers, and high-accuracy micro-positioners—piezoelectric drivers, flexible telescope lenses, or dual-beam tracking techniques—can achieve pointing accuracies within ± 1 microradian, which allow transmission divergences of as much as 3 microradians, well within the specifications of commercial laser transmission systems. Although locating ground stations accurately in positions to minimize atmospheric attenuation and dispersion significantly decreases the pointing risk, it is always possible simply to increase beam divergence—by reducing transmission magnification—and either increasing transmission power or increasing receiver magnification to compensate for actual experienced pointing error.

THE SPACE SEGMENT

The LSC concept is innovatively contrary to both the orthodoxy of GEO fixed satellite service (FSS) and the current fashion of direct broadcast service (DBS) through either GEO or mobile satellites in low earth orbit (LEO) or MEO. Satellite engineers and operators of both first and second generations have become intrigued with, and remain under the spell of, the commercial possibilities of the unbalanced complexity concept. Large, powerful, complex, automatic satellites with many sophisticated, redundant, and fail-safe devices can communicate effectively with simple ground receivers. Rather than facing a small, intensely competitive market of sophisticated ground station proprietors, operators of such high-performance—and very expensive—satellites can sell thousands of small, inexpensive, terrestrial cellular transceivers to a mass consumer market—clients, customers, or subscribers. The trend to complexity imbalance is clear: whereas TELSTAR I weighed 172 pounds and carried a tiny whip antenna that could communicate only with the 85-foot dish antenna at Goldstone, California, modern satellites weigh several tons and can carry thousands of conversations between handheld cellular telephones. First-generation satellites use large, expensive, powerful satellites to broadcast to a large number of simple, inexpensive ground receivers, which may be mobile or stationary. The new second-generation LEO-MEO operators use more, smaller, lighter satellites with intensive digital frequency reuse and signal processing to reduce further the cost and size of ground receivers.

Both approaches require large amounts of transmission capacity in lower RF bands. Both approaches stress heavy investment in the space segment—either a few large, powerful satellites or large numbers of smaller, more sophisticated satellites—coupled with relatively simple ground segments. Laser satellite communication inverts the complexity imbalance by combining a small number of sophisticated ground stations with a small constellation of lightweight, inexpensive satellites in MEO, supplemented by a few more laser-equipped satellites in GEO. Third-generation modulated laser retroreflection technology enables spinstabilized satellites to form an efficient six-satellite constellation of two laser rings in MEO above the Van Allen radiation belts, with a second smaller set of three to five satellites in the GEO. In addition to necessary systems for satellite housekeeping, each satellite carries several lasers equipped with LSC retroreflector subsystems to establish and maintain laser links with ground stations, with other satellites and with spacecraft in one continuous laser ISL. Each laser employs solid-state, phased-array beam steering linked to the satellite's retroreflectors to acquire and track other lasers and to maintain integrity of the laser links. Piezoelectric modulation at 50-100 MHz of flexible tracking-beam retroreflectors at both transmitters and receivers—or even only at the transmitter—determines atmospheric distortion and generates instantaneous compensating distortion of the main beam. As turbulence or aerosols distort the counterdistorted main beam, the result is an undistorted signal beam focused on the receiver. Modulated tracking beam retroreflectors are particularly promising for reducing attenuation in a phased-array beam-steering system. Through up- and downlinks, intersatellite links, and laser links with spacecraft, the LSC satellites receive signals—modulated laser beams carrying message traffic—from earth, from other satellites, and from spacecraft.

The receiving LSC satellite forwards signals to another satellite, or transmits them down to earth for delivery to an addressee. In addition, the uplink to each satellite may be used for beaming additional power to satellites for station keeping, pointing, or operating onboard systems. Coordinated ground-station modulation-demodulation and onboard signal processing determines how a particular message is forwarded through the intersatellite rings and shunted into the proper satellite's downlink with earth. To receive laser-modulated communication traffic from other satellites outside the LSC system, each satellite operates an additional laser retroreflector subsystem linked to a solid state tracking and acquisition system. Guidance and processing signals—or reprogramming instructions—are transmitted on up- and downlinks through the patented retroreflectors on a single beam with traffic being carried.

THE SIGNAL SEGMENT

The signal segment of any LSC system is also in contrast to conventional first- and second-generationapproachesto satellitetelecommunications. Designers of geostationary RF satellite communication systems have historically increased transmission capacity through more transmitters, transponders, and receivers; narrower RF spot beams; higher radiated power; larger apertures and antenna systems; frequency reuse; and frequency coordination to allow collocation of satellites in the GEO. The essence of the LSC signal segment is the simple set of satellite rings that effectively creates laser wired space. Use of modulated laser beams instead of RF broadcast cones to carry signals effectively minimizes the hardware on each satellite, reduces radiated power and interference, increases transmission capacity, and operates through small antenna systems on earth. In addition to high-latitude connectivity, the GEO satellites provide additional capacity and alternative transmission paths. The MEO dual rings offer the low latency and strong signals needed for voice traffic. Two-way RF communications require two receivers and two transmitters, just as orthodox optical communications require two lasers; or lights, pointed at each other.

The LSC modulated laser retroreflector technology uses a single laser coupled with a laser reflector for each laser link both to acquire and track satellites and to receive and transmit information—and also reduces the criticality of satellite attitude control. Dual-frequency laser beams for message and acquisition functions and independent retroreflector modulation based on attitude determination maximize system simplicity, reliability, and capacity, while minimizing weight, volume, and power requirements. The configuration of two offset MEO laser rings with a complementary set of laser GEO satellites allows each satellite to operate either as a repeater or as a laser source. Thus, signals can pass in both directions in the rings, as determined by routing and processing performed in the ground stations. The configuration of nine to ten satellites optimizes the balance between capacity and cost and provides low latency where required, general interconnectivity with other systems, and near-global coverage. Orthodox RF practice places uplinks and downlinks on different frequencies —each of which consumes transmission capacity—and requires a third frequency for acquisition and tracking.

In contrast, the LSC retroreflector system combines these functions in a single beam and doubles capacity with a dual-frequency configuration. Simultaneous initial illumination of satellite retroreflectors with multiple laser beams of different wavelengths and comparison of the datamodulated reflections both multiplies capacity and allows real-time compensation for any atmospheric distortion of laser beams.

Use of the optical—rather than the radio—range of frequencies allows transmission of huge amounts—terabits—of data. The optical wavelengths of laser beams provide capacity beyond that of the entire RF spectrum. Exploiting the capacity of optical wavelengths for telecommunication traffic requires only a technically complex and sophisticated—but eminently practicable, inexpensive, and feasible—process of modulation-demodulation and conversion-reconversion between RF and light. Orthodox RF antenna theory and broadcast technology involve manipulation of antenna size and shape around effective isotropic radiated power (EIRP), aperture, and a large propagation cone, which do not require fine beam pointing. Laser satellite communication, in contrast, uses a highly collimated propagation beam at low power and replaces the large satellite antenna with fine pointing by phased-array beam steering directed by retroreflectors. The ground antenna system can be reduced to a simple 1-meter reflector antenna. To complementhumanity's current fascination with low-capacity, low-speed, low-frequency, high-power broadcast, laser satellite communication provides high capacity and high speed in a high-frequency, low-power, point-to-point communication channel. The LSC system invites migration of high-capacity data and simplex video applications beyond the RF spectrum and exploits the inherent high capacity and high speed of the optical spectrum. The LSC signal segment based on retroreflector laser technology is simple, fast, inexpensive, and reliable.

THE GROUND SEGMENT

For any communication satellite system, the ground segment comprises fixed ground stations, mobile receivers and transmitters, and the satellite control facilities. In addition to the several varieties of mobile receiver-transmitters – airborne, marine, vehicular, and portable – fixed RF ground stations are of four types: high-capacity, mid-capacity, master, and receive-only. High-capacity stations are large, fixed installations that carry trunk communications for multiple large users. Midcapacity stations connect satellites to private networks carrying video conferencing, electronic mail, data transmission, and duplex voice communications.

Master ground stations link very small aperture terminal (VSAT) or ultrasmall aperture terminal (USAT) networks and terminals – carrying digital voice, video, and data – to a satellite. Receive-only ground stations are the least expensive and receive voice-only, data-only, or televisiononly. Like the space segment, the LSC ground segment presents a contrast to the orthodox complexity imbalance of geostationary satellite practice. A few sophisticated ground stations – the minimum is three; a set of five provides redundancy – perform most of the complex functions of operating a communication system through laser links. The interfaces between the satellite system and local service providers in the connected terrestrial telecommunication networks involve several active, intelligent functions that must be performed at each LSC ground station:

- Making physical connections with the terrestrial networks or systems – public or private; circuit-switched or packet-switched; twisted or coaxial copper wire, fibre-optic cable, microwave, RF broadcast, or modulated laser beam – that provide and receive signals; these connections might be any or all of three distinct types:
- For an arriving terrestrial or space laser link carrying a modulated laser beam, the input interface would simply shunt the modulated laser beam into the uplink, while the output interface would simply shunt the downlink into the terrestrial laser link;
- For a terrestrial fibre-optic network or cable connection, the input interface would remodulate the received signal to the uplink laser frequency and shunt it, while the output interface would remodulate the downlink laser signal to the fibre-optic frequency and shunt it;
- For a wireless RE connection, the input interface would remodulate the received RF signal to the uplink laser frequency and shunt it, while the output interface would remodulate the downlink laser signal to the RF broadcast frequency and shunt it;
- Transferring received signals either to laser carrier for transmission to the satellite, or to the appropriate terrestrial carrier;
- Processing signals (modulation and demodulation, multiplexing, amplifying);

- Acquiring and controlling satellites and detecting when a satellite laser beam is degraded below acceptable levels or has disappeared entirely;
- Managing traffic (routing, determining if traffic is in queue during a degradation of the laser beam; determining the nature of traffic for prioritizing and rerouting; storing interrupted or delayed signals, applying appropriate traffic reroute algorithms—store-and-forward, close-hop rerouting, best-network search and forward, cancel, or return to sender—shifting traffic because of congestion; informing senders of errors, delays, rebroadcast, and other carrier states).

The LSC ground stations are most analogous to a multifunctional—highcapacity, mid-capacity, or master—RF ground station, since the LSC system can perform all three functions simultaneously: trunking to multiple users, distribution to separate users, and connection to a discrete network. As for any network of multiple satellites, a LSC network operations centre on earth controls all satellites, couples satellites and ground stations or mobile receiver-transmitters, and monitors system integrity and stability. All of the intelligent functions of the LSC system occur in the ground stations to ensure that the signal arrives and leaves with no degradation, loss, or corruption. Since some satellites are not in geostationary orbit, each ground station—like each satellite—uses retroreflectors to acquire and track each satellite as it comes into range. Since traffic may not be received from the terrestrial network as laser, each ground station remodulates terrestrial signals to laser frequencies and satellite signals to appropriate frequencies for transfer to terrestrial systems.

Each ground station also performs some satellite and network control through a modulated retroreflector link. Intuition suggests the benefits of adapting existing RF ground stations to LSC requirements, much as the common antenna towers, arrays, and fields are able to accommodate different types, frequencies, sizes, and interfaces of various RF antenna systems. Although it may be possible, and clearly cost-effective, in some cases to adapt existing RF ground stations to the LSC system requirements, or even simply to lease space, such factors as location, laws and regulations, interface requirements, and protocols, suggest that such cases will be few.

Location

Sensitivity of laser beams to atmospheric interference—refraction or reflection—is the primary influence on location of ground stations. Lack of cloud cover is the most important criterion. This fundamental condition mandates location of ground stations in areas of low ambient atmospheric aerosols (clouds, airborne pollution, fog, smog, and haze). The ideal stations would be at high altitudes where the weather is dry and clear, regularly spaced around the earth's circumference, but not necessarily on the equator. Some intuitively obvious sites are Australia, Peru, Denver or Phoenix, and southern Spain. Mountaintop locations near astronomical observatories are also ideal. Some of the stations could operate unattended, subject only to scheduled maintenance and emergency repairs. Reliable high-bandwidth telecommunication links to the rest of the world are vital. Fibre-optic cable and terrestrial laser are the optimal media.

Cloud Cover

Distribution of cloud cover over the earth's surface is neither uniform nor even symmetric. Nephanalysis and meteorology have discerned a stable pattern of cloudiness related to latitudinal zones that encircle the earth. In both the northern and southern hemispheres, clouds form preferentially at the mid and low latitudes along all meridians. Minima, with high probabilities of extended clear weather—0 per cent-10 per cent cloudiness—and good seeing conditions—10 per cent-30 per cent cloudiness—occurat the high latitudes and in the subtropics. Within these latitudinal zones, cloudiness is affected by local and regional orological, topological, and atmospheric features. In the Northern

Hemisphere, during all seasons and over oceans and continents, cloud cover is minimal in the Arctic and the subtropics. Over the continents, cloudiness increases with decreasing latitude from a minimum at about 80° N to a maximum between 60° N and 50° N and then decreases to another minimum at 15° N-20° N. The southern hemisphere shows a similar latitudinal distribution of cloud cover: maxima in the low (0°-10° S) and middle (55° S-60° S) latitudes, and minima in the subtropics and Antarctic at all longitudes. Over a year, the average cloudiness of the southern hemisphere is notably lower than cloud cover in the north. Although variations occur cyclically over the seasons, the zones of maxima and minima have remained stable since meteorologists began collecting cloud data in the early twentieth century.

The standard deviation of cloudiness is slightly higher in the Northern Hemisphere than in the south. Whereas Northern Hemisphere cloud cover seems most stable near the equator, in the Southern Hemisphere "minimum variability of f [cloudiness] is observed in the middle latitudes (from 40 to 60° S)." Further, in contrast to a clear latitudinal shift of northern zonal minima with the seasons—because of the northward shift of the intertropicalconvergence(ITC) zone every summer—the southern middle-latitude zonal minimum lies during the entire year between 55° S and 60° S. Daily observations have been combined with broad zonal distribution patterns of cloudiness to develop broad frequencies of cloud cover around an entire global latitude zone. In the northern middle latitudes (35° N-65° N) clear skies appear around the world with a frequency of less than 10 per cent in all seasons. In lower latitudes (15° N-30° N) the frequency of global clear weather is 20 per cent-25 per cent. In all zones of the Northern Hemisphere, global clear skies are the least frequent weather pattern, except in polar zones where clear weather has the highest frequency (30 per cent-45 per cent in summer and 25 per cent-30 per cent in winter). In the Southern Hemisphere, the likelihood of global clear skies in the middle latitudes (45° S-65° S, the monsoon belt) is negligible. In the tropics and subtropics (0°-40° S) global clear skies occur with a frequency of 10 per cent-15 per cent.

The frequency of "good seeing conditions" is greater than 50 per cent in winter in the tropics and throughout the entire year south of about 20° S-25° S—except Antarctica, where at 80° S-85° S the probability of clear skies is 60 per cent-80 per cent in all seasons, but only about 45 per cent-50 per cent at 75° S. Continuing work of the International Satellite Cloud Climatology Project (ISCCP) has refined the latitudinal patterns into a set of probabilities of cloudiness in particular locations for most of the world. Within the broad pattern of latitudinal zones of cloud cover, the ISCCP models identify general locations with low probabilities of cloud cover. Determining definite probabilities for a specified level of cloud cover remains beyond the capabilities of current data, models, and analytical techniques. Although historical details of cloud cover are available for active observation sites, extrapolation of these data to other sites without empirical observations is at best an inexact art of *ceteris paribus*.

Orography

Although the stability of zonal patterns of cloud coverage are sufficiently robust to allow valid generalizations, the clear asymmetry and anisotropy of global patterns of cloud cover demonstrate the powerful effects of local orology, topology, and atmospheric conditions. Beyond the effects of clouds, however, the short wavelengths of laser beams are sensitive to orographic interference. Desert areas generate dust with even slight winds, and vegetative cover around the ground site reduces atmospheric dust. Turbulence—caused by microthermal atmospheric fluctuations—can cause scintillation, beam broadening, spatial incoherence, and phase distortion. The profile of a suitable LSC ground site, thus, includes low cloud cover, predominance of sunshine, low particle scattering, low turbulence, and low ambient humidity. Sites without high probability of clear skies, however, are least suitable for reliable laser earthspace communications. Thus, issues of particle scattering, turbulence, and security

southwest Asia, a broad region of low cloud cover—12 per cent-20 per cent—extends from Turkey through Iran, Afghanistan, and Pakistan, into the Indian Desert southwest of the Himalaya Mountains. Although data are less reliable in central Asia, the Kirghiz Steppe between the Aral Sea and the Tien Shan Mountains on the border of China shows mean cloud cover generally below about 15 per cent-18 per cent. The Siberian Uplands, northern Mongolia, and Chinese Inner Mongolia also show mean annual cloud cover levels of under about 15 per cent. All of Australia north of about 30° S experiences lower than about 15 per cent mean annual cloud cover, with an arc of near-0 per cent cloud cover extending from the Great Sandy Desert in Western Australia through the Sandy Desert in the Northern Territory to the Great Dividing Range in Queensland. In the southwestern Pacific Ocean, a triangular region of low cloud cover—8 per cent-15 per cent—extends from the Caroline Islands—150° E, 8° N—in the northwest to the Tuamotu Archipelago—140° W, 20° S—in the South to about 115° W, 10° S in the northeast—nearly to the Galapagos Islands.

Regulatory Requirements

Although national and international laws and regulations are quite specific on licensing procedures, siting, construction, standards, and operation of RF ground stations, neither the International Telecommunications Union (ITU) nor any other international organization has addressed the quite different regulatory issues involved with laser communication—either terrestrial or satellite. Some national regulatory agencies—but not those agencies regulating telecommunications—have begun to address propagation of laser beams as a hazard, as a form of atmospheric pollution, and as a weapon. No national legislatures or regulatory agencies have considered modulated laser beams as a form of communication. Despite the comprehensive legislative and regulatory treatment of RF waves and propagation, the electromagnetic and physical differences between radio waves and laser beams suggest that existing radio regulatory procedures and processes can be, at best, a loose model for management of laser communications. Until international law addresses the issues of laser communications, laser communication operators must presume that any regulations imposed by host governments on laser ground stations will be *ad hoc*, probably negotiated, controversial, and clearly different from those imposed on RF long-line ground stations.

The major safety concern regarding any LSC system is that of optical injury to anyone who might accidentally wander through the beam. The geometry of the system limits this threat to crew and passengers on aircraft and spacecraft flying through the beam. The U.S. regulations and standards have been established by the American National Standards Institute (ANSI) in ANSI 2136.1, American National Standard for Safe Use of Lasers, and many have been adopted by other countries. For the 810-nm and 850-nm lasers used in the AstroTerra system, the Maximum Permissible Exposure (MPE) is 320 x 10^{-6} W/cm^2. (Since 810 nm is beyond visible range, the critical aversion response time is not applicable.) Since even a low-power 1-watt system constitutes an ANSI Class 4 laser, operators must take appropriate safety precautions at the ground stations.

Interface

The LSC ground segment must accomplish technical integration and ensure compatibility with both public and private terrestrial telecommunication networks, although it can be expected that the signal received and returned to a network will be digital. Any interface between the LSC ground station and a terrestrial network or channel involves at least two considerations: software and protocols; and equipment.

Software and Protocols

The combination of transmission control protocol (TCP) and Internet protocol (IP) (TCP/IP) has become the *de facto* standard for all Internet traffic and for much non-Internet traffic. Although the target markets and applications for any

LSC system are not limited to Internet traffic, the growth of Internet and multimedia traffic suggests that laser satellite communications must incorporate the software and protocols to accommodate TCP, IP, TCP/IP, and TCP over satellite (TCPSAT) standards. Although the features of a particular protocol are transparent to laser technologies, the LSC system collectively, as well as each station individually, must accommodatethe protocols used by the terrestrial networks to which they connect.

Hardware and Equipment

The LSC ground station must feed electrical signals received from a first-generation metal network to a laser exciter coil as a charging pulse to generate a laser burst for transmission. To reconvert a received laser beam to an electrical signal for delivery to a first-generation network, the ground station must direct the received beam through a set of photoelectric diodes and amplifiers to generate electrical signals based on changes in beam intensity. Signals received from a fibre-optic network may usually be shunted directly into the laser transmitter. Although laser satellite communication may need to design specific devices for the particular situation of each ground station, these technologies are commercially available, reliable, and proven. In contrast to RF transmissions, laser beams have extremely low signal-tonoise ratio (SNR). Thus, beyond any controllable errors introduced by lightelectricity conversions, correction of SNR-induced errors are not expected to require the costly, sophisticated error-recovery technologies required for intensive frequency reuse in RF systems. Although laser attenuation is insignificant—even at GEO distance-atmospheric interference and beam diffusion may reduce received energy. Both ground stations and satellites will require self-selecting beam detectors and amplifiers to compensate for energy loss during transmission, which will vary as atmospheric conditions change. Because any amplification, however, introduces additional noise, ground stations will require internal filtering systems to recover the normally low laser SNR.

As with light-electricity converters, these sorts of technologies—amplification and filtration—are commercially available, reliable, and inexpensive. For signal routing, the LSC ground stations must be able to shunt traffic from an incoherent satellite link to a robust link. Technologies and devices for accomplishing the former as a refinement of circuit-switched, or token-ring, routing are commercially available and reliable, although laser satellite communication may need to refine standard devices for specific situations. An alternative approach, where alternative satellites and satellite laser links are available, is packet-switched internal routing. As with the circuit-switched approach, standard, commercial equipment, techniques, and devices are available for packet-switched internal routing. External message routing and recovery from system failure depend specifically on the routing procedures of a connecting terrestrial network. To participate in external routing performed in the connecting terrestrial networks, the ground stations must continuously monitor the entire LSC network to determine whether a received signal can be delivered. In accordance with arrangements with connecting terrestrial networks, the LSC ground station may interface in either of two ways: signal rejection, or delayed transmission—storeand-forward. Using the signal rejection approach, when a ground station identifies the LSC system to a transmitting terrestrial network as nonreceptive, the routing protocols of the transmitting terrestrial network route traffic through other channels, either to another LSC ground station or through traditional RF satellite and fibre-optic cable operators.

This type of routing is appropriate for real-time and high-priority messages. Terrestrial networks—or even individual users-using this approach to routing may establish priorities that the LSC ground station would respect. Using the delayed transmission approach, the LSC ground station holds a received signal until it can be transmitted into the LSC system. This approach —batch mode—is used regularly for transmissions that can be made in batches (e-mail, messaging or paging traffic, data transfer, or trunk video). This approach to routing is most efficient for very large volumes of traffic that require many hours of transmission over RF or cable channels. The devices, technologies, and software for any of these standard routing techniques are commercially available, reliable, reasonably priced, and readily adaptable to laser communication systems.

TECHNICAL ADVANTAGES OVER RF SATELLITE SYSTEMS

The system establishes a dual-coverage global network with only three to five GEO satellites and six MEO satellites. Additional satellites increase flexibility, improve reliability through redundancy, and facilitate expansion of the system by establishment of additional ground stations. The ISL are technically simple, optically stable, and physically reliable. The laser beam operates unhindered by electromagnetic or other interference in the airless vacuum of space. Laser intersatellite links are nearly free from propagation losses. Passive first- and second-generation RF satellites rely on terrestrial networks to interconnect uplinks and downlinks and achieve global coverage through hops with corresponding latency. Intersatellite links obviate all hops and provide global coverage without increased latency. The MEO time-delay for signal transmission (one-tenth of a second) is substantially less than that inherent in RF GEO satellite systems (one-fourth of a second), is not critical for simplex traffic, and accommodates IP modulation forms. The cost of launching and positioning a lightweight satellite in MEO at about 3,500 kilometers (km) above earth is substantially lower—up to 40 per cent-than the cost of putting a large RF satellite into the GEO (35,000 km, or 22,300 miles). With few mechanical components, low power requirements, and small optical and electronic devices, these satellites are both smaller and lighter than conventional high-power, high-capacity GEO satellites.

Early projections of LSC satellite weight are from 200 to 300 kilograms (kg). GEO RF satellites weigh up to 3,800 kg. This light weight allows deployment of the entire MEO constellation from a single launch vehicle. The narrow laser beam focused directly on a single receiver provides high security from unwanted interception, makes jamming expensive and difficult, and allows a laser communicator to control not only the message content, quality, and nature of the signal, but the dissemination of the information that the signal carries.

TECHNICAL DISADVANTAGES OF THE LASER SATELLITE SYSTEM

The MEO satellites are *not* stationary with respect to the ground. They must be acquired and tracked by the ground station through some electro-mechanicaloptical process. The MEO is close to the Van Allen radiation belts. MEO satellites may require some shielding and deteriorate more rapidly than do GEO satellites. Laser uplinks and downlinks are subject to atmospheric interference. Laser transmissions may even lose beam integrity as the beam passes through atmospheric aerosol obstructions (clouds and rain). The laser beam is subject to large propagation losses. Laser beams lose from 0.5 to 1.0 dB in transmission, in contrast to radio wave energy losses of about 0.4 dB between the transmitter and a receiver. A laser satellite system using LSC retroreflector technology cannot broadcast to multiple or mobile receive-only stations. The system is necessarily duplex.

DESIGN AND CONSTRUCTION OF THE SYSTEM

Most of any LSC system's technologies, components, and devices are available in commercial off-the-shelf versions. Design of a prototype would require further research in the areas of solid state steering of laser beams, tracking moving laser sources, and acquiring exogenous laser beams. Construction of an operating space segment would require focused engineering and development of unique devices and components for satellite attitude control, platform stabilization, and laser-based satellite control techniques. In the signal segment, construction of any sort of LSC system will require further research in atmospheric interference, attenuation, and instability of laser beams; development of modulation-demodulation schemes and adaptation of protocols involved in interface with existing telecommunication networks; and determination of the effects of digitization and compression on modulation, demodulation, reliability, or carrier integrity at laser frequencies. An effective LSC ground segment will require further research and development in acquisition and tracking of MEO satellites, interfaces with terrestrial telecommunication systems, and identification of ideal locations of ground stations to optimize the balance of atmospheric conditions and access to terrestrial telecommunication networks.

A BUSINESS MODEL

Although every business deserves its own unique identity, it is possible to speculate about the appearance of any LSC business. From the comfortable perspective of the ideal situation, it is useful—and rewarding—to try to describe its business concepts and strategies, its markets, its structure, its finances, and even to hypothesize about its success.

BUSINESS CONCEPTS

Any business involved in laser satellite communications—to conduct research, operate ground stations, launch and operate satellites, or provide telecommunication capacity to some target market—is a pioneer in the creation of a new industry. Any business concepts will rely on the optical spectrum to alleviate the bandwidth problem commonly associated with RF technology and first- or second-generation satellite and terrestrial systems. Any concept must rely on the LSC intersatellite laser link ring, at least one pair of laser up- and downlinks, and appropriate interconnections or cooperative associations with;

1. A high-speed, high-capacity converged terrestrial fibre-optic network,
2. A FSS system,
3. A teleport operator, or
4. Terrestrial laser or microwave links. Since the LSC operator will own the intellectual property associated with the space segment—and possibly that of the signal and ground segments—an attractive business concept is to lease or license either.

The LSC operator might lease laser bandwidth, much as FSS operators lease RF transponders, to terrestrial users, to FSS or MSS RF operators, or to fibre-optic network operators. Alternatively, LSC might license intellectual property to teleport operators or RF satellite operators to convert existing RF "gateways" to laser gateways. Several complementary operational business concepts, which are not mutually exclusive, are also intuitively apparent for LSC operations:

a. Globalizing a regional fibre-optic network with at least 2 Gbps (OC-48) capacity;
b. Bringing telecommunication service to a developing country with a dry, clear climate;
c. Contracts with governments to augment their strategic RF systems;
d. Satellite news gathering;
e. Carrying simplex traffic from GEO and other overloaded duplex services. Instead of leasing bandwidth to users and operators, the LSC firm might allocate bandwidth to a set of subordinate or associated businesses that sell or lease bandwidth to functionally specialized wholesale customers or clients.

Providing Connectivity

Laser satellite communication systems are well suited for connecting other high-performance networks, especially fibre-optic networks that are physically separated, areas that do not have an extensive terrestrial communication infrastructure, and terrestrial laser systems.

Fibre-optic Networks

Communications are earth-point-to-earth-point with a terrestrial network at each point. The LSC system obtains signals through a node or terminal of a high-speed, high-capacity, converged fibre-optic network —OC-48 operating at 2 Gbps—collocated with the LSC ground station. The system delivers signals to a node or terminal of another network also collocated with the LSC ground station. The network operators obtain signals from original senders, deliver signals to ultimate receivers, and perform retail marketing and sales. A generic prototype is the connection of one converged network or public switched telephone network (PSTN) to another converged system or PSTN in another country, region, or continent. The concept requires cooperative associations with terrestrial fibre-optic network operators at LSC ground stations.

Less Developed Countries

Communications are earth-point-to-earth-point. One point is a node of a rural-remote telephony network in a less developed country (LDC), which is linked to local, national, or regional telecommunication infrastructures. The other point is a terrestrial node of the global telecommunication network. The LSC ground station in the LDC may connect to an existing terrestrial telephone network, a terrestrial laser or microwave network, a local PSTN, or a VSAT or USAT system. One ground station (probably the station in the United States established under another business concept) operates the LSC satellites. The host government, in cooperation with some international or foreign development agencies, develops, funds, constructs, and operates each LSC ground station in each LDC as a development project. The concept requires appropriate associations with the host government, a funding development agency, and terrestrial network operators (fibre-optic, laser, or microwave). This concept has no generic prototype. The LSC system obtains signals from the LDC under the terms and conditions of the development project and provides connectivity through LSC satellites to the rest of the world. Signals could include point-to-point voice, video, and data telephony, but not broadcast transmissions.

A model involving LSC-associated network operators involves installation of local terrestrial networks with the LSC ground station as a node interconnecting local and global networks. A model involving LSC-associated satellite operators involves installation of local VSAT-USAT networks with the LSC ground station as a hub. In either model, each LSC ground station is developed and funded as part of a development project with, through, and for the LDC host government.

Terrestrial Laser Operators

Communications are earth-point-to-earth-point with a terrestrial laser link at one, at least, of those points. The LSC system obtains signals through the terminal of a terrestrial laser link with either a terrestrial network or an ultimate user. The system delivers signals to a node or terminal of another network or to another terrestrial laser link with an ultimate user. The terrestrial laser operator either obtains and delivers signals, or arranges to relay signals through terrestrial networks as appropriate and performs retail marketing and sales. The generic prototype connects news coverage of a disaster or athletic event through a temporary or mobile terrestrial laser link with a regional LSC ground station to a broadcast network's central editing facility.

An alternative prototype for dense industrial and urban environments is a terrestrial laser communications link to relay signals between users and a regional LSC ground station at very high data rates, effectively bypassing the telephone connections—and lower data rates—through copper wire and fibre-optic cable. For both prototypes, a shared —or partial— fibre-optic cable could provide emergency connectivity with the LSC ground station at a much reduced data rate. The concept requires cooperative association with at least one operator of terrestrial communication lasers and possibly with a global news network operator.

Providing Capacity

The inevitable migration of users and applications requires additional capacity and bandwidth at frequencies beyond the RF spectrum. Laser satellite communication is the obvious—and, at the present levels of science, the only-medium that can solve the problem of scarcity of bandwidth at the preferred radio frequencies. Laser satellite communications are eminently able to provide capacity to both users and operators.

RF Satellite Operators

Communications are initially earth-point-to-earthpoint with a fixed satellite service (FSS) or LSC ground station at each point. Replacement FSS satellites are expected eventually to carry laser intersatellite link (ISL) capabilities for space-point-to space-point communications between FSS and laser satellites. The LSC system obtains signals through laser-equipped satellites of FSS systems, or through interconnected FSS and LSC ground stations. The system delivers signals—through either LSC satellites or laserequipped FSS satellites—to some FSS ground station or to a laser-equipped FSS satellite. FSS operators obtain signals from original senders, deliver signals to ultimate receivers, and perform retail marketing and sales. The LSC system could connect to some RF satellite operator's own system either at the RF ground station or aboard the GEO satellite. The ideal structural interface connects the FSS system to the LSC system through signal processing equipment on FSS satellites to shunt signals between onboard RF transponders and laser transmitting and receiving equipment on the same FSS satellite. Establishing laser intersatellite links between RF satellites and LSC satellites would not relieve the RF spectrum pressure on the RF satellite. In contrast, several RF transponders aboard the LSC satellites would allow the RF ground station operators to shunt non-real-time, high-capacity traffic directly into the LSC system by refocusing their RF uplink beams.

At the RF ground station, simple switches and modulators might shunt traffic from the RF uplink, convert RF signals to modulated laser beams, and feed them into a laser uplink to the LSC satellite. A reverse system would return received laser signals to the client ground station through a laser downlink and reverse processing. An alternate structure would connect one FSS ground station to one LSC ground station through terrestrial laser, microwave, or fibre-optic channels, and concentrate signal processing at the LSC ground station. A generic prototype connects one FSS ground station to the LSC ground station in another country. A generic prototype for a future constellation of satellites with ISL between GEO and MEO satellites would connect any FSS and LSC ground station to any other FSS or LSC ground stations serving other countries. This concept requires cooperative associations with FSS operators, and possibly with terrestrial fibre-optic or laser network or circuit operators.

Teleport Operators

Communications are initially earth-point-to-earth-point between a teleport at one point and another teleport or the LSC ground station at the other point. The LSC system obtains signals through a teleport collocated with the LSC ground station or connected to a distant LSC ground station through terrestrial laser, microwave, or fibre-optic channels. The system delivers signals to either another teleport or directly to a terrestrial PSTN operator. A generic prototype of the LSC-teleport relationship is connection of one teleport to another teleport in another country. The concept requires either cooperative associations with teleport operators at LSC ground stations or cooperative associations with FSS operators providing capacity to teleport operators, and may involve cooperative association with terrestrial network operators.

Specialized Operations

The unique features of LSC provide a diffused focus for specialized business operations to meet specific requirements for high-capacity, high-speed, secure, communication channels. Two intuitive such opportunities are the space business and microwave replacement.

Government and Institutional Space Business

Communications are space-point-to-earth-point, earth-point-to-space-point, or even space-point-tospace-point to supplement government civil or military strategic RF communication systems—U.S. Deep Space Network (DSN), Military Strategic Satellite Communications Network (MILSTAR), or Defence Satellite Communication System

(DSCS). The LSC system obtains signals from space exploration missions through an upward-looking laser mounted on the LSC MEO – or FSS GEO – satellites, or through terrestrial or FSS interfaces. The system delivers through the LSC laser ring directly to terrestrial receivers with laser downlinks, or through FSS, terrestrial laser links, or fibre-optic cable, or stores signals and data at some LSC ground station for later retransmission and delivery. The generic prototype is laser connection between a spacecraft and one of the LSC satellites.

The LSC system relays data, sensor information, and imagery from the spacecraft through an appropriate LSC ground station and some terrestrial linkage to the relevant user. The system also relays telemetry signals from the user facility to the spacecraft. The concept requires a dedicated onboard laser with acquisition and tracking capabilities controlled from earth, as well as a laser on the spacecraft.

Adaptation of RF Microwave Infrastructure

Communications are earthpoint-to-earth-point with at least one point connected to a converted or adapted microwave network. The LSC system obtains signals from the connected microwave-laser terrestrial network and from the global telecommunication network. The system delivers signals either to the global telecommunication network or to the laser-microwave network for relay to the ultimate user. The generic prototype is connection of the LSC regional ground station to existing microwave networks that have been converted or adapted to terrestrial laser linkages. The LSC ground station would extend that network through LSC satellites into the global telecommunication network. The concept requires cooperative associations with a terrestrial laser operator and possibly with a terrestrial microwave operator.

TARGET MARKETS

Laser satellite communication provides very-high-capacity, very-high-speed carrier services to several specialized niche markets within the larger market for satellite carriers of telecommunication traffic. Not limited to any single type of traffic—voice, video, or data—LSC is ideally suited to "a customer or market where the overwhelming need is bandwidth, and loss of communication is acceptable. But doesn't put the customer out of business—they simply have

to shift to a less functional system until the primary one returns." The most salient services that these sorts of customers, users, and markets demand are not particular applications, but connectivity to both terrestrial networks and wireless systems in combination with high transmission capacity. Across several broad functional sectors, the physical features of laser beams exclude, however, many real-time or interactive applications. The many visible, functional, consumer applications for telecommunication services—RF or laser, wireless or wired—are the primary markets of the clients, subscribers, users, or customers of any LSC operator. Rather than reflecting particular functional sectors of the vast, fashion-driven telecommunication customer—or retail—markets, the features of the broad LSC or wholesale markets include:

- Connectivity between the LSC ground station and the ultimate user through either terrestrial laser communication links, fibre-optic links—single cables or networks—or some other terrestrial interface, operated by a cooperative associate, strategic partner, client or customer of the LSC operator;
- Long point-to-point communication links; primarily international carriers; large land masses and regions—China, Russia, India, Latin America, Africa—may also be included;
- A substantial amount of long-distance and international Internet, simplex video, and data traffic;
- Data traffic that can tolerate some interruption or delay of service;
- Areas where geography, finance, or politics may interfere with laying land lines —copper or fibre-optic—or providing RF satellite ground stations;

Satellite News Gathering (SNG)

A specialized application for laser satellite communications—in cooperation with an appropriate terrestrial carrier—is provision of high-capacity transmission services to global news gathering organizations between remote news sites and central editorial sites. Many athletic events, concerts, political or other incidents and disaster sites require continuing video and audio coverage. Many such events occur where it is difficult and expensive to lay cables but feasible to install temporary terrestrial laser links. A cooperative associate—or customer—of a laser satellite operator might establish terrestrial laser communication links to relay signals—audio and video—from the site to a local editing station. Permanent fibre-optic or laser connections with a regional LSC ground station would allow subsequent forwarding for transmission to global editing stations or broadcast facilities. Appropriate business associates for any LSC operator in this market might be providers of terrestrial laser communication links or fibre-optic cable operators, as well as a global news gathering organization.

Internet

Modulated laser beams are well suited to provide high-speed, high-capacity carrier services for some specialized sectors of the Internet—especially the 190-230 high-level Internet service providers (ISP) currently using the existing telecommunication networks to carry Internet traffic. Specific Internet applications might include cache data, synchronization, and network database update, electronic mail, network management, research, and voice telephony, as well as others that are yet to emerge. An appropriate cooperative associate for any laser satellite operator in this sector would be an ISP or Internet access provider that also operates an extensive converged, terrestrial fibre-optic network. Laser satellite communications would establish high-capacity global connectivity, which lower capacity RF satellites cannot provide at all and which transoceanic cables can provide only at very high investment cost. With the same business model used by RF satellite operators, the LSC operator might operate either as a wholesaler or as a retailer through a teleport or gateway.

Government, Military, and Scientific Applications

Government and military users across the world require high-capacity, high speed carrier services for routine traffic, high-volume transfers of data and imagery—analog and digital—periodic reports, and other non sensitive, non-real time traffic. A laser satellite communication system is an ideal provider of high capacity, high-speed laser carrier services to scientists exploring deep space —especially NASA—as an alternative and complement to the overloaded NASA Deep Space Network (DSN). As the space-faring countries of the world send more and more probes, missions, and vehicles into space, DSN will be unable to communicate with them as frequently as has been common. Congestion and increasing demand for DSN mission time has forced NASA into the second generation approach of the telecommunication industry: using radio technology to relieve congestion. Beyond the development of new autonomy technologies —to reduce the need to communicate with spacecraft—under NASA's New Millennium programme, the global space exploration community will require additional high-speed, high-capacity transmission and relay channels to meet the communication requirements of accelerating space exploration programmes. Approaches to these market sectors might be either as a prime contractor to government and military agencies or in cooperative association with a major prime contractor.

It is likely that any approach will also involve a cooperative relationship with an operator of a terrestrial fibre-optic cable or laser link. A secondary approach to this market might involve government research projects related to laser communications.

Rural and Remote Areas

Many rural areas are currently served only by microwave telecommunication relay networks or various cellular telephony systems. These microwave networks operate over a substantial infrastructure based on line-of-sight terrestrial links. Microwave relay is being replaced by fibre-optic cable in some areas, but in many regions, terrain and high cost prevent installation of cable and reduce the effective capabilities of the traditional cellular telephony systems. Inexpensive conversion of line-of-sight microwave networks to laser networks extending from a single LSC ground station may be feasible in some situations. The speed with which terrestrial laser communication links can be established suits them well for emergency services in remote areas with LSC connections to global communication networks.

A combination of terrestrial and satellite laser communications is also apposite for the many remote operating sites for oil, gas, and mining, which are within line-of-sight of each other, of the shore, or of prominent terrain features. Where RF microwave communications are difficult—near the horizon, over long distances, because of scatter or Fresnel diffraction—a combination of terrestrial laser and laser satellite communication can provide a very high-speed link with global connectivity. In many areas of active petroleum exploration—the Gulf of Mexico, the U.S. West Coast, Venezuela, and the Arabian Gulf—and in some undeveloped regions—Peru, North and South Africa, South Asia, western China and Mongolia, and interior Australia—combinations of laser and microwave links have experienced availabilities and reliabilities greater than 95 per cent. The most salient users in this sector include exploration companies, government agencies, disaster management firms, nongovernmental organizations, and medical facilities.

Additional Markets

Some additional markets and applications that may offer opportunities to LSC operators include multimedia services, data broadcasting, digital imaging and telephony, alternative or replacement for microwave relay, scientific visualization, and electronic newspapers. Beyond the primary markets, the potential markets and applications for high-speed, high-capacity communication at low cost are limited only by man's imagination.

MANAGEMENT CONCEPT

The third-generation LSC business concept is much more than orthodox diversification. The unique business concept of developing cooperative associations with major firms in several complementary industries suggests adoption, with some refinements, of what has been called the *parenting approach* to management of a diversified business. In addition to managing the functional operations of the firm, the parenting approach includes a primary role for the management team in discovering, developing, and exploiting new opportunities—such as those of third-generation migration, specialization, and global interconnectivity. In a dramatic reversal of the conventional inward focus of management on technology, the parenting approach calls for a predominantly outward focus of managers at all levels on relationships and opportunities.

CORPORATE STRUCTURE

As appropriate for the broad concept extending over several market sectors, technologies, and industries, any LSC operator must establish a central corporate headquarters with a staff section, an operating division, or a focused cooperative association for each of the major business concepts. In addition to the normal corporate functions—general management, administration, finance, legal affairs, contracting, marketing, and business development—corporate headquarters establishes and operates the LSC space segment through a network operations centre. Corporate headquarters conducts both research and development relevant to all aspects of space segment, including design, production, and launch of the satellites; the signal segment (signal conversion and processing, laser modulation and demodulation, software interface, and telecommunication technologies); and the ground segment (siting, hardware interface, and site construction). In conjunction with its corporate associates, corporate headquarters must pursue new business development and opportunities for laser satellite communications.

Cooperative Associations

The preferred structure of the cooperative associations with terrestrial fibreoptic network operators and FSS satellite operators is that of a joint venture between the LSC operating company and another system operator. As a first priority LSC operators should establish cooperative associations with at least one North American converged fibre-optic network operator, at least one foreign fibre-optic network operator, at least one operator of GEO satellites over North America, and at least one operator of GEO satellites over Asia and Europe. These cooperative associations provide laser communications and wireless access servicesthrough high-speed, high-capacitypoint-to-point transmission for satellite traffic that can tolerate delays in transmission—non-real-time.

Corporate Business Divisions

As a second priority, the operating LSC divisions for government, teleports, and institutional business must generate and enter into contracts to provide LSC services and transmission capacity. These contracts not only generate current revenues, but often establish credibility and advantages for subsequent LSC business. Such contracts may require additional arrangements with operators of terrestrial network operators or GEO satellites, which should be integrated into the cooperative associations that are the primary LSC business priorities.

Internal Staff Sections

Lower priority business areas—rural telephony and terrestrial laser—will require at least monitoring and development of potentially productive relationships with organizations and individuals that are active in those areas. Corporate headquarters can maintain an appropriate presence in these areas through internal staff sections. These staff

business concepts focus on the results and success of this initial phase of technology development. The diversified LSC business concept and cross-industry scope of the LSC technologies contrast strongly with the focus of standard capitalization schemes on a specific technology, application, or market. Development of any LSC business, thus, seems inappropriate for traditional first- or second-generation approaches to capitalization. An innovative third-generation approach involves alternative schemes for capitalization, each built around a particular consortium of different types of investors, with different goals, from different industries, reliant on different technologies, and focused on different markets for different applications. Each approach envisions the LSC operator as an autonomous operating business capitalized through four sources of capital—institutional investment, private investors, venture capital, and corporate investment—over several years.

A small initial investment by an institutional investor provides a robust structure for developing an operating consortium among investors from several industries. Small investments by respected and experienced telecommunication executives provide the aggressive, dynamic, innovative leadership that the LSC concept requires and deserves. The venture capital investor exploits the strengths and experience of the venture capital industry in developing emerging technologies.

The LSC Consortium

Each LSC consortium includes at least one major telephone company or terrestrial network operator, at least one satellite operator, a financial institution, and involves a large aerospace vendor. Some also involve many other firms—equipment vendors, software houses, publishers, new networks, and even governments. In this third-generation extension of industrial and financial trends, any LSC entrepreneur should form such a group of cooperative associates from the industries that complement—or perhaps constitute—the global satellite telecommunication industry.

The corporate satellite and network operators are critical operational members of any LSC consortium as both primary investors and strategic allies. Corporate investors are primary prospects for cooperative associations; they also provide an important connection between LSC and the retail customer markets that ultimately generate the demand for capacity. In conjunction with technical development—perhaps under the sponsorship of a venture capital investor—the LSC entrepreneur develops through and with its corporate associates the commercial infrastructures, ground stations, and market presence to operate the business. The combination of four sources of capital—institutional investment, private investment, venture capital, and corporate investment—serves well both to establish the LSC firm as an operating business and to introduce the thirdgeneration LSC concept to a second-generation world. An institutional investor —investment bank—would be most suitable to support the firm's initial efforts to develop a consortium of private, venture, and corporate investors.

At least one individual—private investor—with knowledge, experience, and expertise in the telecommunication satellite industry should be invited to invest a small amount and assume executive or technical responsibilities with the LSC corporate entity. At least one venture capitalist investor should be invited to support development of the technical concepts of the LSC system—space, ground, and signal segments—to test and demonstrate those concepts and to produce and deploy the LSC satellite constellations. The essence of consortium financing lies in the distinct and appropriate cooperative association—joint venture, equity sharing, strategic alliance, or a specific, unique arrangement—with each selected corporate associate. In addition to major investment, the corporate consortium members provide supplemental technological capabilities, internal production capacities, specialized managerial skills, or specific market positions and presence.

Alternative Approaches

Five alternative approaches to business development and capitalization seem intuitively apparent: deployment of the complete GEO-MEO system; deployment of a small GEO system; deployment of a large MEO system reinforced with complementary laser devices installed on other operators' GEO satellites; deployment of a small unreinforced MEO system; and deployment of only a set of ground stations to operate through laser devices installed on other operators' satellites. See Appendix F for an optimistic, speculative projection of semiannual capital requirements for each approach.

The Complete GEO-MEO System

The complete-system approach leads to two interconnected constellations of LSC satellites in GEO and MEO. The most expensive alternative, deployment of the complete system includes a GEO constellation of three to four satellites to support the business concept of providing transmission capacity to FSS operators seeking to migrate traffic to higher frequencies. The MEO constellation of six satellites provides low-latency connectivity between fibre-optic network operators seeking to extend connectivity beyond terrestrial connectivities. The complete system involves four to five LSC ground stations and at least two corporate associates, as well as an institutional investor, a venture capitalist, and several private investors. With adequate capitalization and cooperation, the LSC operator should be able to deploy the complete system within three years of initial capitalization.

The GEO System

Deploying only the small LSC GEO constellation with four or five ground stations is a lower-cost alternative. This approach assumes the ability to install the appropriate signal processing, interface, and laser communication devices on another operator's LEO, MEO, or GEO satellites to establish laser links with LSC GEO satellites and up- and downlinks with LSC ground stations. This approach requires LSC ground stations only where the other operators do not have them, but assumes an ability to shunt LSC signals through other

MSS and FSS systems and to receive signals in space directly from other MSS and FSS systems. Connectivity with other systems may be through intersatellite links, terrestrial connections, or up- and downlinks. Interconnected transmission capacity and speed would necessarily be limited to those of the associated satellite operators. The primary target market for this small GEO system is terrestrial converged network operators seeking global connectivity to other converged networks. The GEO system would involve four or five LSC ground stations, at least one converged network operator as a cooperative associate, an institutional investor, a venture capitalist, and several private investors. As for the complete GEO-MEO system, the LSC operator should be able to deploy the GEO system within three years of initial capitalization.

The Reinforced MEO System

This less expensive alternative assumes the ability to install the appropriate laser communication devices in some modular configuration on another operator's geosynchronous satellites to establish laser links between them and LSC ground stations. Each GEO laser module provides full interconnectivity between the other operator's FSS system and the LSC MEO system. Like the complete GEO-MEO system, a complete version of this approach would also require establishment of at least two ground stations within the range of each laser-equipped GEO satellite and laser intersatellite links between the LSC MEO satellites and the other operator's GEO satellites. A minimal structure might omit intersatellite links and GEO-specific ground stations and operate the LSC MEO system and the laser system carried by the other operator's GEO satellites as separate systems connected only through the LSC ground stations. The Reinforced MEO system involves four or five LSC ground stations and at least two corporate associates—a GEO satellite operator and a converged network operator—as well as an institutional investor, a venture capitalist, and several private investors. The LSC operator should be able to deploy the Reinforced MEO system within three years of initial capitalization.

The Small MEO System.

This inexpensive alternative deploys only the dual laser rings in MEO with the ground stations necessary to operate them. Like the complete GEO-MEO constellation, the small MEO system is not dependent on other satellite systems or operators. Providing low-latency global connectivity, the small MEO constellation of six satellites would focus operations on fibreoptic network operators seeking to extend connectivity beyond continental borders. A complete global system would involve four or five LSC ground stations and at least one corporate associate, as well as an institutional investor, a venture capitalist, and several private investors. With adequate capitalization and cooperation, the LSC operator should be able to deploy the small MEO system within three years of initial capitalization.

The Ground System

Deployment and operation of this initial small groundstation system would demonstrate the technical and commercial feasibility of laser satellite communications, as a robust low-risk foundation for developing and deploying a complete LSC system as a later project. Elimination of the entire space segment obviates the costs of construction and launch of the satellites and reduces the capitalization requirement considerably. Completely dependent on other operators' satellite systems, this approach assumes an ability to install the appropriate signal processing, interface, and laser communication devices on another operator's LEO, MEO, or GEO satellites to establish laser up- and downlinks with LSC ground stations. This approach requires LSC ground stations where they can maintain connectivity with other operators' laser-equipped satellites. The ground-station approach assumes an ability to shunt LSC signals through other MSS and FSS systems and to receive signals directly from other MSS and FSS systems through terrestrial connections, or up- and downlinks. The systems' transmission capacity and speed would necessarily be limited to those of the associated satellite operators. The ground-station system involves four or five LSC ground stations and at least two corporate associates, as well

system. Once they are widely deployed, satellite systems may change the face of the telecommunications industry." The primary physical considerations and constraints on any satellite system are the altitude of the orbit, which determines latency, and the spectrum that the satellite uses, which determines power requirements and capacity to carry information. The three types of satellite orbit are:

- Geosynchronous orbit (GEO) fixed at 22,300 miles; high latency (0.24 sec.); satellites are expensive; power requirements are high; intersatellite spacing of 2° allows only 180 slots in the orbit. Only three satellites can provide geometric global coverage; four or five satellites could maintain continuous global RF coverage with a single satellite covering each point along the earth's equator; six or seven satellites could provide continuous dual coverage; and eight or nine satellites could provide continuous triple-satellite coverage;
- Midlevel earth orbit (MEO) from 6,250 to 13,000 miles; requires several satellites for global coverage; lower latency (0.06-0.14 sec.); mobile satellites cannot be used for telecommunication when over oceans or uninhabited areas;
- Low earth orbit (LEO) at 500-1,500 miles; very low latency (0.03 sec.); requires many satellites for global coverage; mobile satellites cannot be used for telecommunications when over oceans or uninhabited regions. At LEO the numbers of satellites required for global coverage increase to thirty to forty at an altitude of 500 nautical miles.

Since all satellites have historically used RF as the carrier, they could communicate with a single fixed ground stations by using narrow, focused RF beams; or with mobile receivers and dispersed fixed ground stations by using broad broadcast RF beams. Whether broadcasting or beaming, most RF satellites do not operate singly or independently, but as part of a system that involves other satellites: a telecommunication constellation. The cost of installing a constellation depends on the number of satellites involved—usually the number needed to achieve global coverage—the cost of building each satellite, and the cost of delivering each satellite to its proper orbital location and putting it into operation. Satellites destined for fixed satellite services (FSS) roles are the most complicated and expensive to build, with large antenna systems, complex transponders, and high effective isotropic radiated power (EIRP); these GEO satellites are also the most expensive to deploy. Lower satellites can have smaller antenna systems and less powerful transmitters.

They are less massive and less expensive to deploy. The unit costs of satellite manufacture, launch and, deployment rise with orbital altitude. For an entire constellation, however, the intermediate altitudes represent the best compromise between the number of satellites required for global coverage and the cost of building, launching, and deploying each satellite. Constellation cost reaches a minimum at altitudes between 4,000 and 6,000 nautical miles. Although smaller constellations can provide continuous global coverage at higher altitudes, the costs of building and launching each highaltitude satellite are also considerably higher. Low-altitude satellites are less expensive, but coverage of global markets requires many more of them. Compares orders of magnitude for the costs of minimum constellations for global coverage at several orbital altitudes.

Table: Order-of-Magnitude Costs of Satellite Constellations (US$ million)

Orbit*	Launch Cost	Manufacture Cost	Satellite Cost	Constellation Cost
3,000	30	40	70	900
5,000	40	50	90	800
6,000	45	60	105	780
10,000	55	90	145	850
15,000	110	140	250	950
20,000	160	210	370	1,200

The range of satellite telecommunication markets extends from interplanetary communications through global and regional broadcast to trunk television and Internet access to local telephone calls. Satellite services can support most types of communication, but have focused on broadcast, signal relay, trunk connections, telephony, navigation, tracking, imagery, and scientific applications carried at radio frequencies through FSS constellations of GEO satellites. The current, fashionable applications for FSS telecommunications are broadcast to multiple dispersed receivers, video trunking, and voice and data transmissions to mobile receivers. A smaller and growing sector is the market for mobile satellite services (MSS) provided by smaller satellites moving in lower orbits. Historically, the markets for MSS have been small and specialized, although recent technological innovations promise a much larger potential market for wireless, mobile communications. A major market discriminator between FSS and MSS is the difference between the two sorts of services in telecommunication latency. Since the delay associated with a satellite transmission increases with the double of its distance from earth, transmissions from GEO satellites take two to three times as long as those from LEO satellites. This large inherent latency of GEO satellite communications – up to a half-second – preventsany GEO system from providing an acceptable level of performance for such interactive applications as teleconferencing, remote computer operation, telemedicine, or most of the emerging multimedia applications. In contrast, LEO satellites uniformly deliver fibre-optic-like performance because of the short transmission path between sender and receiver and "can be more real-time, interactive."

The International Telecommunications Union (ITU) has developed a recommendation for a satellite transmission latency standard of 400 milliseconds that allows single-hop satellite communication links involving one GEO satellite, but does not accommodate a double-hop through two GEO satellites. Although all single-hop LEO and MEO transmissions meet this ITU standard, the issue of system latency is complicated by terrestrial networks, which also include digital compression codes, switches, signal processors, and tandem path splitters. Even when satellite latency is compliant, the cumulative effect may well be unacceptable to users demanding "real-time conversation or interactivity." About 70 per cent of current satellite users and applications—simplex video and data—do not require real-time service and expect some delay in delivery of transmissions. "An estimated 37 per cent of little LEO satellite market consumers demand real-time service." Similar estimates are not yet available for MEO systems. The several combinations of orbit, ground receivers, and signal latency have created a variety of markets with different technical features and communication requirements. Although are neither discreet nor mutually exclusive, each market is best suited to a particular type of service, which is determined by the technical and operational characteristics of the satellites that provide it.

Table: RF Satellite Telecommunications Market Combinations

	FSS Satellite in GEO	**MSS Satellite in**
	LEO or MEO	
Single	Voice and data	Radio and television
Stationary	telephony	programme
Ground Station	Video transmissions	distribution
(Gateway)	Business services	Paging and messaging
	(private networks,	
	teleports,	
	VSAT-USAT, TV-based,	
	data-based, financial)	
	Paging and messaging	
	News gathering	
Dispersed, networks	Direct audio-video	VSAT and USAT
Stationary	broadcast	
Ground Stations	Business services	
	(VSAT-USAT)	
	Paging and messaging	
Mobile Ground	Navigation services	Paging and messaging
Stations	Fleet tracking	Cellular telephony
	Paging and messaging	

The long list of retail RF service providers—satellites in LEO, MEO, and GEO, as well as terrestrial cellular and broadcast systems—continues to grow as computer and digital technologies focus on consumer applications. As markets demand more and more bandwidth, limitations on RF capacity in L-, P-, C-, and Ku-band—and most recently Ka-band—are increasing congestion. The exploding success of first-generation commercial satellite services and the high expectations of second-generation operators are forcing development of ways to share these available RF bands and use them more efficiently. Laser satellite communication technologies have been demonstrated to be commercially feasible, although no laser satellite communication systems have yet been deployed. Better wireless-cable interconnection nodes and refinement of archaic regulatory environments are enhancing the possibilities of fibre-optic-RF—and eventually laser—systems as holistic approaches to telecommunications for the future.

FIXED SATELLITE SERVICES

Systems providing FSS are deployed in the GEO. Fixed satellite RF services are best suited to one-way, simplex broadcast of radio and television from a single point to multiple receivers. Another preferred application is providing high interconnectivity among a large number of users. Geostationary systems are best suited to providing regional, rather than international networks. Although fixed RF satellites can also carry high-density telecommunication trunks, high digital capacity and quality make fibre-optic cable the preferred transmission route where it is available. Appendix C identifies current GEO FSS operators. Although FSS will continue indefinitely, the FSS portion of the total satellite RF telecommunication traffic is decreasing. The RF portion of FSS will gradually contract to broadcast as other FSS traffic migrates to laser, low-latency MSS, or fibre optics. For economic reasons, increased use of fibre-optic cable for high-density traffic is beginning to erode the dominance of satellites as a highdensity trunk connector.

As business and private networks increasingly adopt less expensive very small aperture terminal (VSAT) or ultrasmall aperture terminal (USAT) technologies and structures, the role of geosynchronous FSS may focus on radio and television broadcast distribution, mobile networking, navigational services, and rural or remote services. "Only 18 per cent of the commercial telecommunication satellites to be launched within the next six years—some 160 of about 885—are heading for GEO. Over 60 of these systems are already operating. Of the GEO satellites to be launched through 2004, 39 per cent are offering traditional telecommunication and television-broadcast services. Direct-to-home TV-broadcasting satellites represent 28 per cent; mobile systems, 15 per cent; multi-purpose spacecraft 15 per cent; and the smaller, multi-media satellites, about 3 per cent." The parallel structures of the market and the industry for fixed satellite services reflect six broad functional types of services: voice and data telephony, bulk video transmissions, direct broadcast, multimedia applications, digital imagery, and business services, which may include any of the other functions.

GEOSTATIONARY VOICE AND DATA TELEPHONY

Every country's public telephone organizations (PTO) use satellites as trunk carriers for international and intercontinental telephony. Another use is providing national connectivity in countries with poor wired infrastructure. The satellites carrying this traffic are operated by the national PTOs or the international satellite organizations (ISOs)—the International Telecommunications Satellite Organization (INTELSAT), the European Telecommunication Satellite Organization (EUTELSAT) ArabSat, and the International Maritime Satellite Organization (INMARSAT). Since a transponder aboard a GEO satellite can generate from $1.5 million to more than $4 million annual revenues according to type and amount of traffic, many commercial GEO operators—PanAmSat, Loral, or GE—also carry some telephony traffic. In Europe and North America, an extensive terrestrial infrastructure and coverage by fibre-optic cable that is beginning to offer multimedia services, the popularity of cellular telephones, and high densities of telephones offer few opportunities for satellite telephony. Satellites have historically provided the primary trunk services for international telephony, but transoceanic fibre-optic cables are reducing the demand for satellite capacity. In 1996, transoceanic cables carried 64 per cent of transatlantic telephony traffic and voice paths, whereas satellites carried only 14.5 per cent of all international telephone traffic to and from the United States.

By 1998, the U.S. Federal Communications Commission (FCC) estimated that only about 3 per cent of all U.S. long-distance calls were relayed by satellite. In some contrast to Europe and North America, the Asia-Pacific region has broadly undeveloped terrestrial and cable infrastructures. With low densities of telephones and large distances, the Asia-Pacific region is the world's fastest growing telephony market. The growing popularity of cellular telephones, as well as large investments in cable networks, however, suggest that satellite carriage of voice telephony circuits in Asia-Pacific may fall from 32 per cent in 1994 to about 20 per cent in 2001, while satellite carriage of data telephony may rise slightly from 17 per cent in 1994 to about 20 per cent in 2001. The 1997-1998 series of Asian currency crises interrupted this continuing market trend. After strong growth since 1994, Asian-Pacific space-segment revenues—voice, data, and video traffic—grew only marginally in 1997 to about $2.6 billion through increased usage of satellite networks by current users. "However, in the second half of 1997, transponder orders, specifically for Ku-band transponders, were down due to the onset of the currency crisis."

Throughout the Asia-Pacific region, the primary factors creating demand for satellite telecommunications arise from factors unrelated to the financial crises, which balance the expanding effects of cellular telephones. An increasing requirement for FSS transponder capacity for both fixed and mobile communications reflects pervasive Asian "telecommunications modernization, decreasing transponder lease or sale prices, demand for the Internet and the creation of intranets, and the need for wide area coverage." By about 2001, the region will recover strong growth rates, which are expected to continue through at least 2004.

Conventional Remote and Mobile Telephony

Although cellular systems and fibre-optic cable networks are commercially attractive in urban areas, fixed wire or short-range cellular services are often impossible in remote, rural areas due to terrain, cost, poverty, and political instability. The dramatic increase in demand for satellite systems optimized for land mobile services or remote, dispersed telephony is especially strong in the developing countries of the Middle East, Africa, Asia, and Latin America. Some of this demand is to supplement existing cellular and microwave relay systems, but most is arising in the huge rural, remote markets for basic telephone services. Direct voice communication through a satellite from individual mobile terrestrial transceivers is provided through several systems. Canada relies on a GEO system for affordable, briefcase satellite telephones. Most of the world turns to INMARSAT-M or Mini-M for navigation aids and maritime telecommunications. For most of the developing countries, any sort of basic telephony services are provided through noncommercial satellite systems—INTELSAT, ArabSat, INSAT, EUTELSAT, EtiSalat. In addition to the separate, rapidly growing market for business services, the market for mobile or remote telephony has three broad, overlapping sectors: remote mobile service, messaging services, and rural or remote area telephone service.

Commercial responses to this demand have been many and varied—GEO, MEO, and three sorts of LEO systems—but most have targeted affluent markets for specialized mobile services, rather than the huge—but probably unprofitable—demand for basic telephony. The noncommercial sorts of remote, mobile telephony, which use GEO orbiting satellites, are being joined by commercial LEO satellite systems including Iridium, Teledesic, GlobalStar, Odyssey, and data-only

OrbComm with their handheld transceivers. The $1-$6/ minute cost of these systems' sophisticated services may limit the market, however, to traveling business executives and will certainly limit penetration of less affluent markets for basic telephone services in developing countries.

Like terrestrial messaging services over a public switched telephone network (PSTN) or private fibre-optic network, satellite messaging services include storeand-forward communications, electronic data interchange, e-mail—public Internet and proprietary services—voice and telefacsimile mail. Satellite systems can usually deliver messages to remote areas within a few hours, at costs of $0.001-$0.01 per byte of data—$1-$10 per minute for voice. Currently dominated by GEO systems—INTELSAT, INMARSAT, AMSC, TMI, and PanAmSat—by 2001 several new LEO systems expect to begin claiming market share in this still-not-well-served market. Remote mobile service extends cellular telephony to areas not well served by terrestrial connectivity. Providing rural and remote telephone and data service to areas not served by terrestrial networks is especially demanded in developing countries. With initial projected tariffs of $1.00-$1.50 per minute, TRW Odyssey and Ellipso, but also Teledesic and Aries, are responding to this demand. This market's "fivefold to sixfold growth in six to seven years could be among the highest in the entire wireless industry. Certainly the ability of the new low and medium earth orbit satellite systems being able to provide high-quality voice and data service with low latency are key service features."

VSAT Rural and Remote Telephony

In addition to their prominence in business networks, some very small aperture terminal (VSAT)-satellite overlays provide basic public telephony in remote areas—especially Asian-Pacific, although VSAT is beginning to penetrate Latin America and Africa also. In both the advanced markets of Europe and North America and the developing regions of the world, business VSAT users have historically emphasized corporate trunking and point-to-point applications as "a viable, if not a more cost-effective alternative to leased line technology." The VSAT satellite network operators "act as middlemen between the corporate users and the satellite operator." The cost for installing most VSAT systems is from $2,000 (urban) to $10,000 (remote) for a terminal [LEO systems project terminal costs of several hundred to $3,000]. Whereas an urban leased line—copper coaxial—may cost $150-$250 per month, a firm with rural facilities faces per-circuit costs in leased-line networks of more than $1,000 per month. "Using VSAT, the same company will have to pay only about $365 a month for a nationwide coverage.... onetime cost is only 25 per cent of operational costs."

GENERATION

The process and function of telecommunications in human affairs involve the complex interactions of science and technology, economics and politics; producing and using complex equipment and providing sophisticated services; government policy, standards, and regulation; as well as markets, investments, and competition. The nontechnical soft aspects, especially, are continually changing and often highly political: information is rarely neat, unambiguous, or even provable. Differences in perspective, focus, national interests, clashes between standards-makers and policymakers—as well as technology itself—can make results unpredictable. Despite the rigor of the physics behind it, the human enterprise of telecommunications is not rational. Neither public policy nor commercial practice are optimized around unambiguous paradigms or algorithms, or even into efficient systems. Telecommunication facts are not always as they seem; conventional wisdom is often unwise and occasionally wrong.

The short history of human telecommunication is not merely prologue to future versions of the past. Momentous generational shifts have occurred, and another is about to occur. Within generations, forecasts and projections are simply linear extensions into the future. But across generational shifts, as Arthur C. Clarke has recognized, "The future is not what it used to be." "Someday the networks that connect our telephones and our computers may indeed use light from end to end instead of electricity, beginning a new Information Age in which communications capacity seems infinite and free." The maturing second-generation telecommunication industry and market have already begun the shift to optics. Driven by the wave of demand for Internet services, deregulation, and innovations in fibre-optic technologies, growth in investment in optical infrastructure and equipment for terrestrial networks is more than double the parallel increases in spending on electrical telecommunication equipment—about 8 per cent per year. The demand for high-speed, high-capacity telecommunications has nearly overwhelmed the venerable switched telephone networks that have served so well for so long. Only a new generation of telecommunications—equipment, infrastructure, channels, media-can meet it. Only optics—fibre-optic cables on the ground and lasers in space-can fill that need.

Laser satellite communication is the third generation of the human telecommunication enterprise. The very high frequencies, huge capacities to carry information, and potential bandwidth of modulated laser beams make them attractive carriers for future telecommunication systems. Conventional wavelength multiplexing techniques efficiently and simultaneously transfer multiple wide-band channels at laser frequencies. Dense wavelength division multiplexing over laser carriers optimizes the transmit and receive systems to encode, decode, and route traffic to many multiplexed users through a single transceiver system. Modulated laser light carrying digital and analog information to, through, and from a constellation of third-generation laser satellites "can support real-time interconnections between any two points on the globe at data rates of several gigabits per second and higher,... using smaller, lighter antennas (telescopes) with comparable or lower power requirements than [those] used in current radio frequency (RF) systems operating at lower data rates."

The birth of telecommunication—long distance communications—in the nineteenth century introduced innovation, invention, science, discovery, and hard work to communications, put messages through wires, and bounced radio waves across the oceans. Telecommunications changed the human experience, expanded human life, and generated a new modern human culture. As the twentieth century advanced, a maturing first-generation telecommunication industry relayed radio signals through huge towers and airplanes and wired the world for telephones. Science and industry learned to use radio frequencies to relay signals through manmade satellites to bring telecommunications to every corner of the globe. The first-generation telecommunication satellites were deployed in geosynchronous earth orbit (GEO) at 22,300 kilometers (km) (about 36,000 miles) above the earth. Since GEO satellites appear from earth to be stationary, or fixed, in the same spot in the sky each day, they are said to provide fixed satellite services (FSS). These satellites with their associated terrestrial terminals or nodes operate as "satellite networks, which are space/ground systems performing a specific function at the appropriate frequencies via a designated orbital slot." The large GEO satellites receive RF signals from earth, re-amplify and retransmit them—usually on a lower frequency—back to earth either as a broadcast RF cone or a focused RF beam.

Earth stations receive the retransmitted signals, process them, and direct the processed signals into a terrestrial network, or directly to the intended end user. The market and the related industry built around providing these fixed satellite services dominate global telecommunications. A postwar second generation of telecommunication engineers converted old science and new concepts into a veritable flood of new technologies. The secondgeneration built heroic economic infrastructures and elegant political institutions — like those in other industries—to bring efficiency and order to telecommunications. A second generation of mobile satellites in lower orbits promises a broad new range of mobile satellite services (MSS) provided by satellites that are not in the GEO and, thus, appear to be mobile in the sky. These second-generation satellites operate in low-earth orbit (LEO)—400-1,000 miles above the earth, in midlevel earth orbit (MEO), or in high elliptical orbit (HEO). Although mobile satellites have been in use for several years, the first operating mobile telecommunication satellite networks will be operating in 1999 or 2000. The combination of telecommunication technology with computer science led to the creation of the Internet—a global network of computer networks linked through various telecommunication links. Telecommunication operators—terrestrial, wireless and eventually in space—operated multiple alternative protocols, implemented competing systems that were either clumsy or clever, and offered multiple solutions—each less than optimal, and often not compatible.

A first-generation monopoly of terrestrial networks evolved into a second generation of economic and technical competition between wire and wireless. The respective industries and an aggressive generation of telecommunication entrepreneurs shifted investments from infrastructure to technology and made technology the driver of the market. The market brought competition, demand, and congestion as operators struggled to use ever more intense and clever technologies to overcome the inherent physical—and imposed regulatory-constraints of wires and radio in meeting what still seems to be insatiable human demand for telecommunication capacity. As congestion in telecommunication channels appeared and increased, innovations in radio and fibre-optic technologies competed for market share—each promising more bandwidth, the capacity and speed of a telecommunication medium to transmit information. Spurred as much by competition and the market as by scientific curiosity, communications engineers learned that shorter wavelengths carry more information, use smaller antenna structures, and allow sharper transmitted beams. Ever since Marconi invented radio, communications engineers have been extending this basic wireless technology to higher and higher frequencies—radio engineers developed long-wave radio, then short-wave, then microwave, and telephone engineers learned to transmit light through optical fibres.

Light has been used as a carrier for decades in terrestrial wire-carried applications, most notably the globe-spanning fibre-optic communications networks upon which the major telephone companies depend. The invention of the free-space laser made possible the use of light as a carrier for wireless communications. In the last decade, the U.S. government began to investigate the possibilities of laser-carried communications between ground stations, aircraft, and satellites. [7] Research and experiments – scientific, industrial, and government – have confirmed several advantages of laser over lower frequencies for point-to-point communications. Transmitters and receivers are smaller and lighter for a specified distance; laser requires lower power for a given distance and data rate; laser can carry higher data rates for a given distance and power; and laser provides higher security and greater resistance to interference.To relieve the congestion strangling second-generation systems, a nascent third generation – terrestrial fibre optics and laser beams in space – embodies concepts that can make holistic telecommunications and universal connectivity both possible and affordable. The very large capacities of the optical spectra invite migration of particular telecommunication applications to frequencies and transmission channels best suited to them, in priorities arranged by market forces. The inherent technical compatibility of laser beams and fibre-optic cables brings seamless global telecommunications comfortably into view as a practical objective for both science and industry. Like the concept of multimode global fibre-optic networks, laser satellite communications brings together the fundamental developments that are rendering even the most advanced concepts of the second generation obsolescent: global deregulation, frequency reuse, optical technologies, and the Internet. Open-market deregulation throughout the world allows and encourages the market to extend telecommunications to the farthest limits of humanity. Frequency reuse in its many forms increases the efficiency with which radio frequencies can carry information, even though it cannot increase the physical capacity of those frequencies.